FROM
SCIENCE FICTION
TO
SCIENCE FACT

Dedicated to Anne Hooper in celebration of nearly
70 years of sci-fi fandom, and gratitude for a
lifetime of imaginative inspiration.

Author Acknowledgements
Thanks to Alison Moss and Guy Croton.

This is an André Deutsch book

Text © Joel Levy 2019
Design © André Deutsch Limited, 2019

First published in 2019 by André Deutsch
A division of the Carlton Publishing Group
20 Mortimer Street
London, W1T 3JW

Printed in Dubai.

A CIP catalogue for this book is available from the British Library.

ISBN: 978-0-233-00609-3

FROM SCIENCE FICTION TO SCIENCE FACT

HOW WRITERS OF THE PAST INVENTED OUR PRESENT

JOEL LEVY

ANDRE
DEUTSCH

CONTENTS

INTRODUCTION

The British writer and historian of science fiction, J. G. Ballard, believed that reading science fiction ought to be compulsory.

He described it as "the most important fiction that has been written for the last 100 years", and that was nearly 50 years ago. For Ballard, sci-fi did not simply explore novel concepts or entertain, it actually helped to create the future. "Everything is becoming science fiction," he wrote, "From the margins of an almost invisible literature has sprung the intact reality of the 20th century. What the writers of modern science fiction invent today, you and I will do tomorrow."

From the vantage point of the twenty-first century, how accurate is Ballard's contention? Some of the most central tropes of sci-fi, or at least sci-fi in the popular cultural imagination, stubbornly resist becoming reality: flying cars, personal jet packs and robot butlers are the most commonly cited failures of sci-fi prediction. Yet there is a myriad of vital and ubiquitous technologies that science fiction not only predicted, but in many cases helped to bring into being. The iconic futurist artist and designer Syd Mead once described science fiction as "reality ahead of schedule".

This book explores the visions of the writers, futurists and far-sighted inventors who made those realities, from the direct influence of H. G. Wells on the atomic bomb and the tank, to the ambitious prototypes created by inventors ahead of their time, such as Nikola Tesla's remote-controlled drone ships. The history and development of each technology is detailed and related in context, exploring the road from prescient fictional representation to real-life technology. In these pages, you will meet the greatest names and works in sci-fi, from Jules Verne and Aldous Huxley to Arthur C. Clarke and Isaac Asimov, *Star Trek* to *The Six Million Dollar Man*, alongside visionary inventors such as Tesla and Wernher von Braun.

LEFT The Amphicar was launched in 1961 but failed to catch on. Among its drawbacks, a strategic application of grease was required after each immersion.

OPPOSITE Poster for the 1936 film *Things to Come*, written by H. G. Wells and based on his 1933 book *The Shape of Things to Come*.

PART 1
MILITARY

01

BURST UPON THE WORLD:
H.G. WELLS AND THE CREATION OF THE ATOM BOMB

The atomic bomb that detonated on 16 July 1945 in the New Mexico desert was the culmination of a colossal scientific and engineering programme known as the Manhattan Project.

Widely regarded as one of the greatest feats of collective endeavour in history, the Manhattan Project corralled the intellectual, physical, military and economic power of the world's mightiest nation to achieve an unparalleled unity of purpose. Yet all this incredible labour and expenditure was predicted in a work of fiction: a fantasy cooked up by the science fiction author H. G. Wells 32 years earlier. In his novel *The World Set Free*, Wells describes "atomic bombs which science burst upon the world". From where exactly did Wells draw his inspiration, and how did, he in turn, inspire the creation of real atomic bombs that science would "burst upon the world" as little as three decades later?

THE CRACK OF DOOM

In fact, H. G. Wells was not the first novelist to imagine that the immense potential of atomic energy could be used for ends of unparalleled destructiveness. In 1895, the Irish journalist and speculative fiction author Robert Cromie published *The Crack of Doom*, in which a crazed demagogue named Brande plans to vaporize the Earth and possibly the entire solar system with "the vast stores of etheric energy locked up in the huge atomic warehouse of this planet". Elsewhere in the novel, Brande calculates that "one grain of matter contains sufficient energy … to raise a hundred thousand tons nearly two miles". His evil plot is thwarted when the formula for his "disintegrating agent" is altered by

the hero of the novel, and the subsequent explosion is limited to destroying the island where Brande is based. Presciently, it is a South Pacific island, not too dissimilar to those that would later be battered by American, British and French fusion bomb tests.

Cromie's novel was written before the discovery of radioactivity, which came a year later when Henri Becquerel showed that uranium salts produced some form of radiation that could fog photographic plates. It would be a further two years after that until Marie and Pierre Curie showed that the source of radioactivity is atomic rather than chemical. Yet there were hints of a new atomic physics emerging; in the same year that *The Crack of Doom* was published, X-rays were discovered by Wilhelm Roentgen (see Chapter 12), working with apparatus devised by the physicist William Crookes that could produce a stream or ray from the cathode in an electrical circuit. These cathode rays were widely suspected to consist of sub-atomic particles, although this would not be confirmed until 1897. So Cromie may have drawn inspiration from then current thinking in physics, although he was also influenced by the mystical pseudo-science of theosophy, hence his use of terms such as "etheric energy".

FURIOUS RADIATION

A writer with a firmer grounding in science who had followed with keen interest the late nineteenth-century discoveries around radioactivity was the aforementioned English author Herbert George Wells. His novel *The World Set Free: A Story of Mankind*, serialized in a magazine from December 1913 and published as a

novel in 1914, details a devastating world war in 1958 and the world government that develops afterwards. The reason that the war is so destructive and has such a shocking effect on global politics is that it involves the use of nuclear weapons: "atomic bombs, the new bombs that would continue to explode indefinitely and which no one so far had ever seen in action." These bombs are powered by Carolinum, a radioactive substance that generates "a furious radiation of energy [that] nothing could arrest".

Wells was writing almost a decade after Einstein's celebrated paper of 1905, *"Does the Inertia of a Body Depend Upon Its Energy Content?"*, which introduced to the world the equation $E = mc^2$ and the concept of "matter-energy equivalence". This means that matter and energy are two sides of the same coin; they are inter-convertible, with matter effectively being a highly concentrated form of energy. In Einstein's equation E is energy, m is mass and c is the speed of light in a vacuum. Since the latter is of colossal magnitude, the equation shows that a tiny amount of mass can be converted into an enormous amount of energy.

Wells goes on to give a detailed, albeit obscure, account of the design and mechanism of these bombs:

> Those used by the Allies were lumps of pure Carolinum, painted on the outside with unoxidized cydonator inducive enclosed hermetically in a case of membranium. A little celluloid stud between the handles by which the bomb was lifted was arranged so as to be easily torn off and admit air to the inducive, which at once became active and set up radio-activity

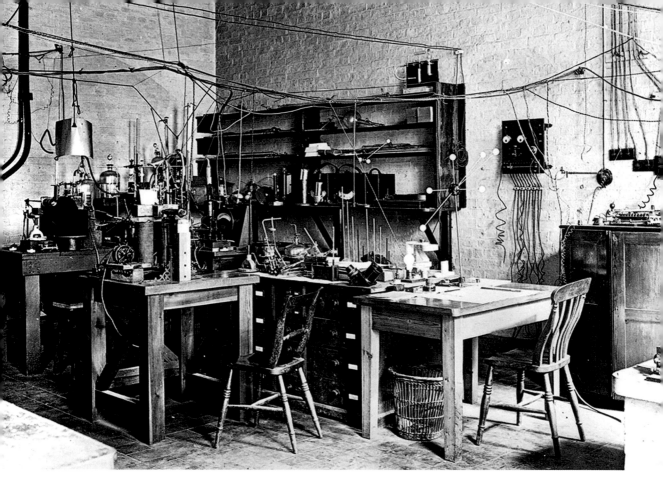

in the outer layer of the Carolinum sphere. This liberated fresh inducive, and so in a few minutes the whole bomb was a blazing continual explosion.

Concealed beneath the largely invented jargon is a description of a nuclear chain reaction, in which radioactive decay of a lone atom triggers the decay of other atoms. It was this key element that would prove to be such an inspiration in the quest to make the fictional atomic bomb a reality.

WAR MADE MORE IMPORTANT
Part of Wells's prediction seemed to be coming true within months of his book being published, as the "war to end all wars" – the First World War – broke out and raged across the globe. Even as war reached its industrialized apotheosis, work continued on the science of the atom.

The leading light in this field was New Zealand-born physicist Ernest Rutherford. In 1909 his team had revealed the secret structure of the atom, showing that it consisted of a heavy, dense, positively charged nucleus surrounded by negatively charged electrons. By 1918, Rutherford's research had begun to raise incredible possibilities. When he was told off for being late to a committee meeting, he responded, "Speak softly, please. I have been engaged in experiments which suggest that the atom can be artificially disintegrated. If this is true, it is of far greater importance than a war." But as Bernard and Fawn Brodie pointed out in their history of weapons *From Crossbow to H-Bomb*, "His discoveries could not be more important than a war because they [had] made war so much more important."

How much more important started to become clear in 1932, when John Cockcroft and Ernest Walton, working under Rutherford at the Cavendish Laboratory in

12

ABOVE Ernest Rutherford flanked by Ernest Walton and John Cockroft, who had just split the atom.

OPPOSITE Rutherford's laboratory at the Cavendish Laboratory at Cambridge University, where the first steps were taken towards atomic power.

Cambridge, became the first people to "split the atom". The scientists used a new particle accelerator to fire protons at lithium atoms, achieving transmutation of the elements and liberating some energy. The *New York Times* trumpeted that "Science has obtained conclusive proof from recent experiments that the innermost citadel of matter, the nucleus of the atom, can be smashed, yielding tremendous amounts of energy". However, most of the science community was far less hyperbolic, their beliefs summed up by Rutherford's insistence that "we cannot control atomic energy to an extent which would be of any value commercially, and I believe we are not likely ever to be able to do so ... Our interest in the matter is purely scientific". The general consensus among scientists around this time was that the internal bonds that hold the nucleus of the atom together are so strong that there would be no way to alter them in order to release meaningful quantities of energy.

However, H. G. Wells's 1913 novel inspired the Hungarian physicist Leo Szilard to think otherwise. Its description of a chain reaction set him thinking about another discovery made by Rutherford's team at the Cavendish: in 1932, British physicist James Chadwick had confirmed the existence of a sub-atomic particle that Rutherford had predicted – the neutron. Unlike the proton, the neutron has no charge, and so it can approach the positively charged nucleus and be absorbed into it. Szilard knew that, at the atomic scale, ejection from the nucleus of a neutron is accompanied by the release of a huge amount of energy. Individually such events might not register on the macro scale, but Szilard theorized that if it were possible to generate a cascade of neutron emissions, an explosive amount of energy could be released. In October 1933 he wrote, "a chain reaction might be set up if an element could be found that would emit two neutrons when it swallows one neutron."

NUCLEAR FISSION

At the time it was not clear if the phenomenon Szilard was describing would be possible, so his contention received little attention. However, subsequent discoveries in 1938 would change that state of affairs dramatically. The products of bombardment of uranium atoms with neutrons unexpectedly turned out to be two much lighter elements: German scientist Lise Meitner and her nephew Otto Frisch realized that the uranium nuclei must be destabilized by absorbing the neutrons, so that they split into two like an amoeba dividing by fission. In 1939, the pair published a paper about nuclear fission, noting that it releases a tremendous amount of energy. However, it was Szilard who saw that the more important issue – in terms of weaponizing the phenomenon – is whether chain reaction of neutron emissions can result from fission. By this time Szilard had been driven out of Europe by the Nazis and had taken up residence at Columbia University in the United States, where his research showed that fission can result in a neutron release chain reaction. Szilard recognized the gravity of his discovery and began to warn fellow researchers that it would be prudent to keep their results secret.

Szilard and the many *emigré* nuclear scientists like him were soon to have their own representations in fiction. In 1936, British author Eric Ambler's novel *The Dark Frontier* featured an atomic scientist who carries a deadly atomic bomb secret formula. He is driven into exile by the Nazis, triggering a race to stop the formula from falling into the wrong hands. Ambler's notion of the weapon itself is sketchy at best, but the plot device signalled growing levels of cultural anxiety about the possibility of science fiction super-weapons becoming science fact.

As Szilard's Wells-inspired speculations began to make this possibility a reality, official anxieties about the transmission of ideas in the opposite direction began to mount. In 1939, Szilard and others recruited Albert Einstein to write to President Roosevelt, in order to alert him to the possibility of atomic weapons and the peril of Nazi progress in this field. The financier Alexander Sachs delivered the letter to Roosevelt in person, capturing his attention with a small slice of alternative historical fiction. Sachs related the (probably fictional) tale of how Napoleon had rejected a scientist who suggested developing a fleet of steamships, a research project that could have turned a global conflict and changed history. The story worked and Roosevelt famously declared, "This requires action!" Work on what was initially known as the "Uranium Project" began in July 1940, and by the end of 1942 the Italian-American scientist Enrico Fermi had overseen the successful construction of the first nuclear reactor, underneath the bleachers of the University of Chicago football stadium.

Fermi's reactor took the form of a "pile" of mixed uranium isotopes. Isotopes are variants of an element that differ in the make-up of their nuclear nuclei; some are much less stable and therefore more radioactive than others. The Uranium Project had already determined that for a bomb, it would be necessary to collect a critical mass of at least 2 kilograms (4.4 pounds) of the uranium isotope U-235, but this isotope accounts for less than seven per cent of naturally occurring uranium. To isolate and purify the requisite U-235 would require a colossal engineering effort on an industrial scale. This would

ABOVE Leo Szilard, the physicist who was so alarmed by Wells's predictions about atomic weapons that he urged others to take the threat seriously.

OPPOSITE The letter to President Roosevelt, written by Einstein at the request of Szilard and other concerned scientists, which helped to set in motion the Manhattan Project.

Albert Einstein
Old Grove Rd.
Nassau Point
Peconic, Long Island

August 2nd, 1939

F.D. Roosevelt,
President of the United States,
White House
Washington, D.C.

Sir:

Some recent work by E.Fermi and L. Szilard, which has been communicated to me in manuscript, leads me to expect that the element uranium may be turned into a new and important source of energy in the immediate future. Certain aspects of the situation which has arisen seem to call for watchfulness and, if necessary, quick action on the part of the Administration. I believe therefore that it is my duty to bring to your attention the following facts and recommendations:

In the course of the last four months it has been made probable - through the work of Joliot in France as well as Fermi and Szilard in America - that it may become possible to set up a nuclear chain reaction in a large mass of uranium, by which vast amounts of power and large quantities of new radium-like elements would be generated. Now it appears almost certain that this could be achieved in the immediate future.

This new phenomenon would also lead to the construction of bombs, and it is conceivable - though much less certain - that extremely powerful bombs of a new type may thus be constructed. A single bomb of this type, carried by boat and exploded in a port, might very well destroy the whole port together with some of the surrounding territory. However, such bombs might very well prove to be too heavy for transportation by air.

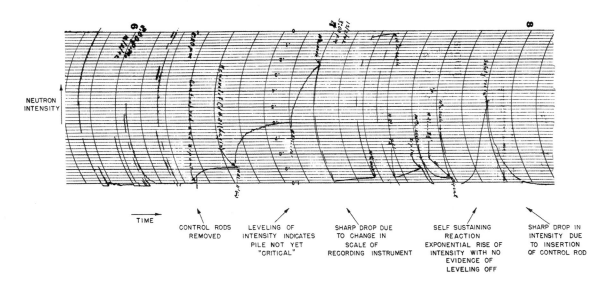

DEC. 2 1942 START-UP
OF
FIRST SELF-SUSTAINING CHAIN REACTION
NEUTRON INTENSITY IN THE PILE AS RECORDED BY A GALVANOMETER

NEUTRON INTENSITY

TIME

CONTROL RODS REMOVED

LEVELING OF INTENSITY INDICATES PILE NOT YET "CRITICAL"

SHARP DROP DUE TO CHANGE IN SCALE OF RECORDING INSTRUMENT

SELF SUSTAINING REACTION EXPONENTIAL RISE OF INTENSITY WITH NO EVIDENCE OF LEVELING OFF

SHARP DROP IN INTENSITY DUE TO INSERTION OF CONTROL ROD

become one of the major goals of what, in 1942, was renamed the Manhattan Project. At vast sites at Oak Ridge in Tennessee and Hanford in Washington, plants and nuclear reactors were constructed in order to enrich uranium ore and transmute uranium into plutonium (for a slightly different bomb design). These projects employed thousands of scientists and workers and there was considerable concern about information leaks and the problems inherent in keeping such an immense endeavour secret.

HIGHLY VITAL PRECAUTIONS

To cover projects such as Manhattan, along with many other areas of national security, the United States instituted a wide-ranging censorship regime with a set of "Codes of Wartime Practices for the American Press". These codes included the "request that nothing be published … about 'new or secret military weapons … [or] experiments.'" Guidance issued to "Editors and Broadcasters" in 1943 by the Director of Censorship noted that, "In extension of this highly vital precaution,

ABOVE Trace recording release of neutrons as a measure of the intensity of nuclear fission during the historic first operation of Fermi's pile.

OPPOSITE Fermi's pile, in which nuclear power was for the first time unleashed and then controlled by means of "control rods" such as the one shown being inserted.

you are asked not to publish … any information whatever regarding … Production or utilization of atom smashing, atomic energy, atomic fission, atomic splitting, or any of their equivalents." The guidance went on to give an exhaustive list of radioactive substances that were not to be mentioned, including uranium, thorium and deuterium.

The language of the guidance clearly shows that concepts of atomic energy were common currency at the time, and indeed atomic energy, including atomic weapons, had featured in science fiction multiple times since H. G. Wells's *World Set Free*.

In 1930, for instance, Charles W. Diffin's story "The Power and the Glory" specified thorium as the source of nuclear power that could be weaponized, while Harold Nicolson's 1932 novel *Public Faces* featured the politics and diplomacy around the invention of an atom bomb. Donald Wandrei's 1934 story "Colossus" featured an ominously inverted prophetic conceit, with Japan threatening to drop devastating "bombs of a new nature" on her enemies to bring a rapid end to a conflict.

Atomic energy and weapons were a particular hobby horse of John W. Campbell, editor of the seminal American "pulp" science-fiction magazine *Astounding* (by 1938 renamed *Astounding Science-Fiction*), who held a degree in physics and was a forceful exponent of stressing the "sci" in "sci-fi". Campbell's first published sci-fi story, "When the Atoms Failed", featured conversion of matter into energy. At his prompting, and often with his technical guidance, a series of atomic energy and weapon tales appeared in *Astounding* magazine in the early 1940s. Robert Heinlein's 1940 story "Blowups

Happen" concerns the stresses and perils of nuclear power, while his 1941 "Solution Unsatisfactory" (written under the name Anson MacDonald) concerns a radioactive weapon and is startlingly prescient in passages. A character in the story explains how his team had been "searching … for a way to use U235 in a controlled explosion", detailing their "vision of a … bomb that would be a whole air raid in itself, a single explosion that would flatten an entire industrial center". For good measure Heinlein went on to predict intercontinental ballistic missiles as a suitable delivery system for such a device.

THE LEGEND OF "DEADLINE"

The most celebrated example of Campbell-prompted atom-bomb prediction came with Cleve Cartmill's 1944 story "Deadline", the reaction to which has become one of the great legends of science fiction. Cartmill's story, the premise and details of which were given to him by Campbell, concerns an alien power that seeks to make a nuclear fission bomb, with particular attention paid to the problem of isolating and purifying the fissionable uranium isotope. "They get U-235 out of uranium ores by new atomic isotope separation methods ... They could end the war overnight with controlled U-235 bombs", the tale's narrator explains.

Sequestered in Los Alamos, the top-secret research facility in New Mexico, the nuclear scientists working on the Manhattan Project were startled to read this story in the pulp journal of which several were avid fans. They had believed their project was shrouded in secrecy, yet Cartmill's story, though inaccurate in the details of its actual bomb, seemed uncannily well informed about the greatest technical challenge facing the Manhattan Project – isotope separation. Physicist Edward Teller (who later

OPPOSITE Striking cover art for a postwar issue of Campbell's influential sci-fi magazine, with a tale about a post-atomic apocalypse society.

BELOW Workers leaving the Oak Ridge facility; most were kept in the dark about the top-secret project on which they were working.

helped create the hydrogen bomb), recalled that the story "provoked astonishment in the lunch table discussions", and that security officers immediately took note.

Campbell relished what happened next, as officers from the Counter Intelligence Corps visited him in his

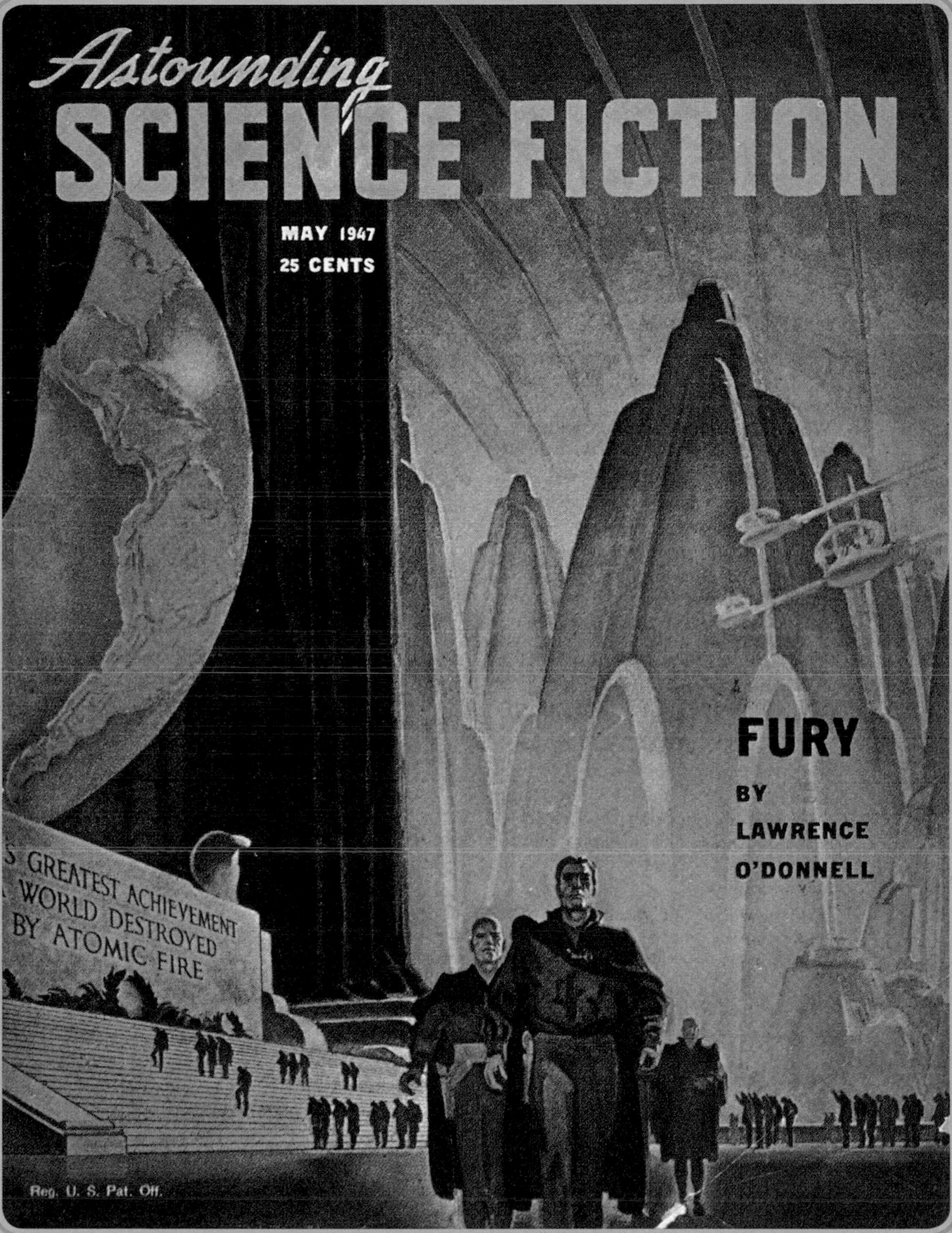

Astounding SCIENCE FICTION

MAY 1947
25 CENTS

'S GREATEST ACHIEVEMENT
WORLD DESTROYED
BY ATOMIC FIRE

FURY

BY
LAWRENCE
O'DONNELL

Reg. U. S. Pat. Off.

office and quizzed him on the story. How had Cartmill got hold of apparently classified information? Campbell pointed out that all the technical aspects of the story were drawn from publicly available sources such as science journals, which he had combined with his own background expertise. Cartmill's role had been purely as a sort of amanuensis, commissioned to write up Campbell's detailed outline. In the pulp editor's telling of the tale, what the intelligence officers failed to notice was a map of subscribers pinned to Campbell's noticeboard, with a tell-tale cluster of pins around an obscure location in New Mexico.

Just how close had Campbell and Cartmill come to revealing the greatest secret of the war? In practice the details of the device in "Deadline" were inaccurate, and as Campbell had pointed out to the intelligence officers, any astute observer of trends in nuclear physics would have been aware of issues such as the challenge of separating enough fissile uranium isotope to create a critical mass capable of sustaining a fission chain-reaction. More generally, well-informed observers may have known that something like the Manhattan Project was probably underway. Isaac Asimov, a science fiction writer who, during the war, worked alongside Robert Heinlein in the research institute at the Philadelphia Navy Yard, claimed to have worked out for himself that such an effort was in progress.

It is also notable that Campbell and Cartmill were well aware that their tale might stir up official interest. In 1942 Campbell himself had written an editorial entitled "Too Good at Guessing", in which he vowed that *Astounding* would steer clear of near-future fictions for fear of giving away something that might breach censorship codes. He seemed to have changed his mind when he commissioned Cartmill in August 1963, but both men were obviously aware that what they were discussing might raise censorship issues. Cartmill wrote to Campbell expressing concern that "there is the possible danger of actually suggesting a means of action which might be employed", but he was apparently ignorant of the censorship guidance that specifically prohibited stories about atomic

ABOVE Workers' housing at Los Alamos, a previously obscure desert village that abruptly acquired a high concentration of sci-fi fans.

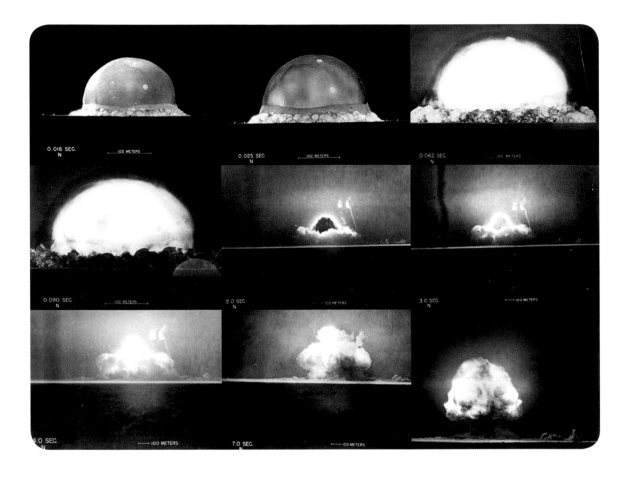

ABOVE Photographic sequence of the Trinity test, the first nuclear bomb explosion, on 16 July 1945.

OPPOSITE The ruins of Hiroshima, after the first atomic bomb was dropped on it on 6 August 1945.

weapons, and Campbell did not enlighten him. The editor's only concession to censorship was to advise that the story be set on an alien world, prompting a rather lazy effort from Cartmill, who made the antagonists of his narrative the evil Sixa powers and the noble Seilla.

The records kept by the intelligence officers who investigated the "Deadline" incident reveal their conclusions that there had been no nefarious intent and no espionage or leaks. But they also show that one of their concerns was that the sci-fi story might cause a leak of information the other way: *into* the Manhattan Project rather than out of it. One of the key tools for maintaining the confidentiality of the project was compartmentalization, so that individuals working on it would know about only their specific field and only a few at the very top would have an overview of the entire enterprise. The fear was that a story such as "Deadline" might be read by one of these compartmentalized workers and help them to fit the pieces together and break out of their compartment.

After Hiroshima the world faced many new questions and challenges. Some of these had already been anticipated by science fiction. In Heinlein's "Blowups Happen", for example, the moon is lifeless because of a nuclear apocalypse. Meanwhile, in *Reply Paid*, a 1942 story by Gerald Heard, the asteroid belt is the remnants of a planet destroyed by nuclear catastrophe. Lester del Ray's story "Nerves", published in *Astounding* magazine in 1942, was a prescient treatment of the threat of an accident at a nuclear power plant. Nuclear conflict became a default plot point for much post-apocalyptic fiction. Fictional treatments of such a scenario, from Raymond Briggs's *When the Wind Blows* (1982) to the TV dramas *The Day After* (1983) and *Threads* (1984), played an important role in shaping public perception of the risks of nuclear war and helped stoke anti-nuclear sentiment.

02

TANKS:
WELLS'S LAND IRONCLADS TO THE MARK IV

H. G. Wells's attentiveness to the technological plausibility of the elements of his fiction – and to current trends in science and society – lent particular power to his futurology.

This meant that his fictions seemed to have a heightened chance of becoming fact. Just as his fictional atomic bomb had helped to inspire its real-life counterpart, so there seemed to be a direct lineage running between the real life tank of the First World War and Wells's 1903 prediction of a similar vehicle. This, at least, was the contention of Winston Churchill, a key figure in the development of the tank, who in 1925 testified to a Royal Commission that it was Wells's story "The Land Ironclads" that had dreamed the tank into existence. However, Wells was not the first to conceive of an armoured vehicle of this ilk; the fictional antecedents of the tank run all the way back to the Renaissance.

DA VINCI'S ARMOURED WAR CARRIAGE
It may seem contentious to refer to Leonardo da Vinci as a science-fiction author, but many of his futuristic designs were purely imaginary, since they were not practical or realizable and in many cases were not intended to be. During Da Vinci's lifetime, Renaissance Italy was the pre-eminent theatre of war in Europe and rival princes competed in the development of military technology, as well as in art and culture. Da Vinci was fully engaged in this cultural-military enterprise, filling his notebooks with dozens of designs for war machines of varying degrees of fantasticality, from bizarre crossbows and machine guns to armoured warships that foreshadowed the technology of a Bond villain. In 1485 Da Vinci was

in the service of Ludovico il Moro – "The Moor", Duke of Milan – and among the ideas he worked up for presentation to his employer was a kind of covered war chariot or carriage, often described as an armoured car or precursor of the tank. "I shall make covered chariots, that are safe and cannot be assaulted," Da Vinci noted, "cars which fear no great numbers when breaking through the ranks of the enemy and its artillery".

The most iconic feature of the design is its conical covering, which protects the operatives within. Sources differ as to the material for this cover: some say it was wooden, which would limit its utility as armour; others that the wood was to be reinforced with metal plates, which would make the contraption extremely – probably impossibly – heavy. Da Vinci was here drawing on classical archetypes, notably the ancient Roman army's tortoise formation. This comprised of a squad of legionnaires bunched tightly together, with those at the fringes holding their shields close together, while those in the centre raised their shields over their heads to make an impenetrable shell from which spears projected. Da Vinci's sci-fi update of the tortoise had the shield wall mounted on a load-bearing chassis, sitting on axles with wheels to carry it around the battlefield. Light cannons were arranged around the skirt of the cover, to provide 360-degree firepower.

Da Vinci drew a complex arrangement of gears as a mechanism within the carriage so that the wheels could be turned by a team of eight men. He considered using animals for motive power but dismissed the idea, recognizing that they would become unmanageable in the confines and din of the carriage. Indeed the whole device would have been unmanageable, partly because it would have been too heavy, and partly because of a flaw in the mechanism as drawn by Da Vinci, which meant that turning the cogwheels would make the front and back wheels turn in opposite directions. It has been suggested that this was a deliberate mistake, included as a sort of intellectual watermark that would booby-trap pirates trying to steal his idea, or even that it showed Da Vinci to be an undercover pacifist, self-sabotaging his own invention.

At the top of the conical cover was a lookout post, from which a commander could survey the battlefield and direct the carriage, a feature with obvious similarities to twentieth-century tanks. Da Vinci's invention extended to his vision for the potential tactical impact of such advanced weaponry, foreseeing that mobile armour could screen the movement of lighter units. "Behind [the armoured carriages]," he wrote, "the infantrymen shall follow, without fearing injury or other impediments".

WELLS'S LANDSHIPS

Da Vinci's armoured car never got further than his drawings, and arguably was never intended to be much more than a fantasy. A practical realization of the principles Da Vinci had elucidated would not come for over 400 years, preceded by another fictional articulation, this time by H. G. Wells. In his 1903 story "The Land Ironclads", published in the *Strand Magazine*, Wells drew inspiration from the ironclad warships that were revolutionizing naval technology and sparking an arms race between Britain and Germany amidst widespread speculation about the imminence of a conflict between the great powers. In Wells's story, he imagined a future war between two

BELOW Da Vinci's 1485 drawing of a "covered chariot" bristling with cannons and protected by its armoured skirt.

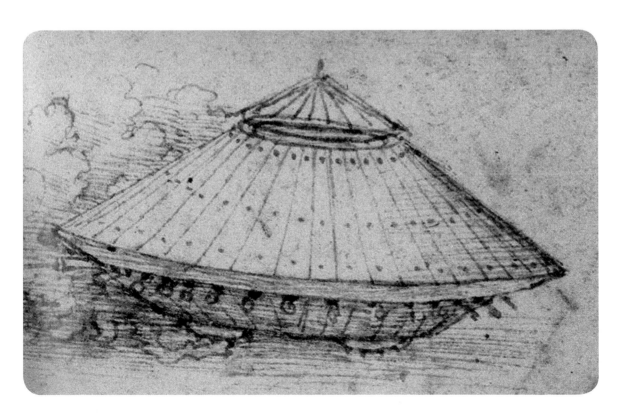

ABOVE Detail from an illustration of Wells's 1903 story "The Land Ironclads", showing the titular behemoths advancing with menace.

nations, which quickly degenerates into a trench-war stalemate. This in itself was clear-sighted of Wells, who was abreast of recent developments in military practice. The Second Boer War, the primary inspiration for the combatants in his tale, had featured trench warfare, and late nineteenth-century conflicts, from the American Civil War to the Franco-Prussian War and beyond, had hinted at the growing dominance of defence over attack and the fearful toll that rapid fire could inflict on advancing soldiers. Such developments had transformed the battlefield and tacticians were struggling to come to terms with them. Wells's brilliant insight was that mobile armour could be the key to unlocking the defensive stalemate. In his tale this innovation would take the form of great armoured landships.

The titular Land Ironclads are 14 steam-powered, 30-metre (100-foot) long behemoths, covered with armoured "turtle" shells that extend in skirts almost to ground level – remarkably similar to Da Vinci's design, but extended lengthwise. In a passage describing the first clear sight of the vehicle, the narrator sees it "in the bleak grey dawn, lying obliquely upon the slope and on the very lip of the foremost trench, the suggestion of a stranded vessel was very great indeed". He estimates its length to be between 24 and 30 metres (80–100 feet), and its height to be around 3 metres (10 feet). Its armour is "smooth ... with a complex patterning under the eaves of its flattish turtle cover. This patterning was a close interlacing of portholes, rifle barrels, and telescope tubes – sham and real – indistinguishable one from the other."

ELEPHANT FEET AND CATGUT

Wells went into great detail about some ingenious technologies incorporated in the landship. The armament consisted of automatic rifles, a technology that was just beginning to be explored in prototype at the time of Wells's writing. These "ejected their cartridges and loaded again from a magazine each time they fired, until the ammunition store was at an end". They had rather elaborate but nonetheless ingenious aiming and electronic-firing mechanisms. A sight "threw a bright little camera-obscura picture into the light-tight box in which the rifleman sat below", whereupon the rifleman selected his target using a dividers-like device, and then pressed a button "like an electric bell-push" to fire. Use of catgut wires allowed for an early form of adaptive optical targeting by naturally compensating for humidity levels that in turn affected "the clearness of the atmosphere".

Correctly surmising that the muddy morass of the modern battlefield would pose a major challenge for mobile armour, Wells had his landships heaving their bulk about on mechanical feet that "were thick, stumpy … between knobs and buttons in shape – flat, broad things, reminding one of the feet of elephants or the legs of caterpillars". Extraordinarily, these feet were a real invention that had only that year been cooked up by the engineer Bramah Joseph Diplock; he called them "pedrails". In Diplock's scheme, the feet were arranged around a wheel and could be used on heavy vehicles to improve their ability to cope with broken terrain. By 1910 he had fixed a "chaintrack" around the outside of the wheel to create an early type of caterpillar track. Wells had each landship move about on eight pairs of pedrail wheel, each on an independent axle, powered by compact steam engines that could propel the vehicles at over 10 kilometres per hour (6 miles per hour). The motion of the vehicle was directed by a commander who could climb up into a retractable conning tower, similar to a submarine periscope.

Although Wells does not specify the armour plating of the landship, he does mention that the adjustable skirt has a thickness of 30 centimetres (12 inches), suggesting that the fixed armour would be at least as thick. In practice such massive armour would make the vehicle prohibitively heavy and unable to cross soft ground. In Wells's fictional world, however, the Land Ironclads turn the tide of war and allow the invading forces to break through the trenches of the defenders, smashing open a breach that is then exploited by cavalry and bicycle corps. Just 14 of the behemoths conquer the entire enemy army.

FROM LANDSHIP TO TANK

In Churchill's version of the tale, Wells and his story were the key inspiration for military developments in 1915. Churchill had been a fan and friend of Wells for well over a decade by this time. They had first met in 1902, a year after Churchill had written to Wells in response to receiving a copy of his new book *Anticipations*, gushing that "I read everything you write". The pair corresponded for the rest of Wells's life, and Churchill would borrow ideas and even phrases from the science-fiction writer, most famously the title of his 1948 book *The Gathering Storm*, the first volume of his history of the Second World War. In 1931, Churchill said that he thought he could "pass an exam" in the work of H. G. Wells.

Churchill's familiarity with Wells's oeuvre evidently extended to the tale of the Ironclads. In 1915, Churchill, as First Lord of the Admiralty, became the driving force behind a naval research project to develop armoured landships on caterpillar tracks, based in part on proposals by Colonel Ernest Swinton. He did so in the teeth of vociferous opposition; the initial proposals had been panned by the War Office, with Lord Kitchener objecting that such landships would simply become targets for enemy artillery: "the armoured caterpillar would be shot up by guns". Interestingly, Wells had foreseen similar objections; in his story the introduction of the ironclads takes the defenders by surprise so that they are not able to get their heavy guns into position quickly enough. A similar strategy would later become one of the key elements in making real tanks effective. What Churchill perhaps appreciated better than less

imaginative colleagues who were not sci-fi fans is that the whole point of the landship is to provide armour mobile enough to avoid being targeted by heavy guns. Churchill had to overcome scepticism even within his own ministry, with the Fourth Sea Lord ranting: "Caterpillar landships are idiotic and useless. Nobody has asked for them and nobody wants them."

Nevertheless, under Churchill's auspices, caterpillar landships began to take shape. An initial prototype nicknamed "Little Willie" had a caterpillar-tracked chassis bearing an armoured box. This was called a "tank" in order to help conceal the true nature of the

OPPOSITE TOP Little Willie, the first tank, was a prototype to demonstrate the combination of armour and caterpillar tracks.

OPPOSITE BOTTOM Big Willie shows off its moves during testing in early 1916.

ABOVE A British tank in action at the Battle of Cambrai, November 1917, the first engagement in which tanks proved their worth.

research project. Little Willie was superseded by "Big Willie", a prototype with a distinctive rhomboidal shape intended to help it navigate broken terrain, bridge trenches and make it hard to overturn, just as Wells had envisaged. By early 1916, Big Willie had impressed military observers sufficiently to be commissioned into service as "His Majesty's Landship, Tank Mk I", and the Mk I would first see action in September 1916, during the Battle of the Somme. An eyewitness description by a young signal officer named Bert Chaney could easily have come from Wells's story:

> …lumbering slowly towards us came three huge mechanical monsters… Big metal things they were, with two sets of caterpillar wheels that went right round the body. There was a huge bulge on each side with a door in the bulging part, and machine guns on swivels poked out from either side.

Despite their evident psychological impact, this initial deployment of the tanks did not go well. In spite of some occasional successes (such as the Battle of Cambrai in November 1917), the early tanks did not have the decisive, war-ending battlefield impact of which Wells had dreamed. However, at Amiens on 8 August 1918, 450 tanks smashed through the German lines and helped the British Fourth Army capture 28,000 men and 400 guns, inflicting a defeat that General Erich Ludendorff declared "in the history of the war, the German army's Black Day". From then on, tanks would become one of the most important military assets and engender an entirely new philosophy of armoured warfare.

MANY INVENTORS
To what extent was Churchill justified in claiming that Wells had been the first to foresee the tank, as he said in sworn testimony to a Royal Commission hearing after the war? As detailed above, mobile armour had been

conceived of at least as far back as Da Vinci's time, and in fact there had been many other precursors in the intervening years. For example, in 1855 James Cowen, a military veteran involved in healthcare, took out a patent on a steam-powered armoured landship he described as a "land battery", which he later revised into a more sophisticated device known as "The Devastator". In the 1880s, French artist and sci-fi

ABOVE Simms's Motor Scout armoured quad motorbike, fitted with a Maxim machine gun, c.1899.

OPPOSITE Cowen's locomotive "land battery", known as "The Devastator". Note the axle scythes, reminiscent of ancient chariots.

writer Albert Robida described future warfare in a trilogy of works about the twentieth century, imagining both giant and small mobile armoured landships. Robida even foreshadowed Wells's notion that armed cyclists might accompany such landships. Around the turn of the century, German-English engineer F. R. Simms cooked up a range of armoured vehicles, from the War Scout – an armoured quad bike – to the Motor War Car and the armoured *draisne* (a kind of railway cart). In 1904, the French firm Charron, Girardot and Voigt built a set of turreted armoured cars for the Russians.

However, it was Ernest Swinton who was the true pioneer of the real-life tank, for it was he who put

OPPOSITE Patent drawing of 1913 for a caterpillar-tracked tractor, which would go on to become the basis for the Holt motor tractor.

BELOW Charron, Girardot and Voigt armoured car, one of the first armoured land vehicles to see service. Some were still in action at the start of the First World War.

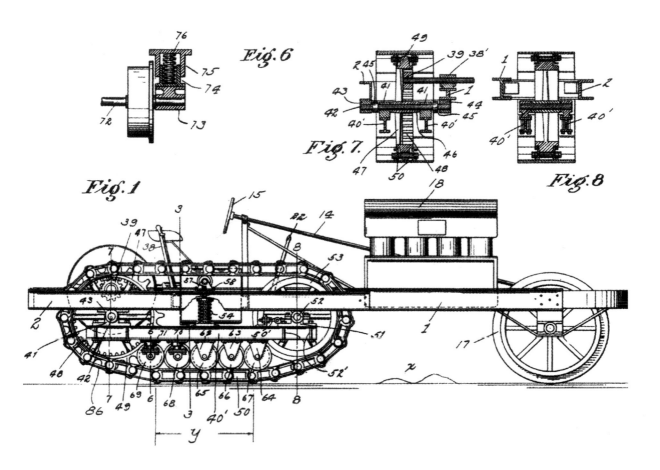

Fig.6

Fig.7

Fig.8

Fig.1

together existing technologies in a real-world context. The caterpillar track – an essential tool for negotiating broken and muddy ground – had been invented back in 1801. Swinton saw it in operation in October 1914, on American Holt motor tractors being used to haul heavy loads across difficult terrain. He realized that this could be a key technology in developing modern siege weapons, of the type needed to break the deadlock of the siege warfare into which the Western Front had quickly degenerated. Moreover, Swinton was only one of several who had advocated various types of armoured, tracked vehicles. In this sense, Churchill was mistaken in celebrating H. G. Wells as the inventor of the tank.

Yet perhaps if Churchill had not read Wells's story, he might not have had the vision that made him such a driving force in the realization of the tank. As early as January 1915, Churchill had written to the prime minister to advocate for armoured steam tractors that could roll over barbed wire, cross trenches and shield the infantry, his enthusiasm inspired by his faith in Wells's powers of prognostication. When the War Office passed on the landship idea on 17 February 1915, Churchill was primed to take over; the Admiralty's Landship Committee met on 22 February. It was this committee that would set the tank project in motion, and make the fictional landships of H. G. Wells become a reality.

03

ENERGY WEAPONS:
FROM THE MARTIAN HEAT RAY TO LASERS AND "ACTIVE DENIAL" MICROWAVE WEAPONS

The laser gun is probably the quintessential science-fiction weapon – standard issue for everyone from Buck Rogers and Dan Dare to Luke Skywalker and Captain Kirk.

It is even in danger of becoming a battlefield reality, with tactical lasers finally on the brink of deployment by active units in real-world conflict. However, lasers are only one of a suite of similar weapons long heralded by sci-fi and sought after by real-life military researchers. These include heat rays, microwave lasers (aka "masers"), particle cannons, plasma rifles and electric charge weapons. They can all be subsumed under the heading "directed-energy weapons". Where conventional weapons fire solid projectiles and/or explosives, directed energy weapons use pure energy or very tiny, highly energetic particles such as ions (atoms with electrons knocked off so that they have electric charges), or ionized gas known as "plasma". Their theoretical benefits include speed-of-light travel, pinpoint accuracy over very long distances and devastating destructive power, as well as infinite ammunition and low cost per shot.

THE HEAT RAY

Probably the best-known example of the death ray in early science fiction is the Martian "heat ray" that features in H. G. Wells's classic novel *The War of the Worlds*, published in 1897. In a dramatic and still shocking passage in chapter five, humanity first encounters this awful weapon when three men bravely go into the

impact crater caused by a Martian projectile in order to make contact with the aliens. The men disappear from the narrator's line of sight, and moments later "there was a flash of light, and a quantity of luminous greenish smoke came out of the pit in three distinct puffs, which drove up, one after the other, straight into the still air". A machine then rears up from the crater and "the ghost of a beam of light seemed to flicker out from it". As the beam swings round, anything in its path flashes brightly and bursts into flame.

The following chapter opens with an attempt at an explanation for the terrible power of the Martian weapon:

> It is still a matter of wonder how the Martians are able to slay men so swiftly and so silently. Many think that in some way they are able to generate an intense heat in a chamber of practically absolute non-conductivity. This intense heat they project in a parallel beam against any object they choose, by means of a polished parabolic mirror of unknown composition, much as the parabolic mirror of a lighthouse projects a beam of light. But no one has absolutely proved these

OPPOSITE Cover art by Edward Gorey for a 1960 edition of *The War of the Worlds*, showing Martian tripods roaming the English countryside.

THE WAR OF THE WORLDS

H. G. WELLS

LOOKING GLASS LIBRARY

details. However it is done, it is certain that a beam of heat is the essence of the matter. Heat, and invisible, instead of visible, light. Whatever is combustible flashes into flame at its touch, lead runs like water, it softens iron, cracks and melts glass, and when it falls upon water, incontinently that explodes into steam.

Wells makes a technical error in describing the ray as a "beam of heat": heat is an energy flow with associated entropy (a kind of dissipation of energy into other forms such as vibration of atoms), and thus cannot be projected very far or with much coherence. What Wells describes is basically a laser; a coherent beam of light that effectively carries no heat and so can transfer very high

ABOVE Death rays blaze in all directions across the dramatic poster for the 1953 film adaptation of *The War of the Worlds*.

OPPOSITE An engraving from the seventeenth century showing the principles of Archimedes' burning mirror, and its effect.

energies to whatever surface it falls upon. Thus, as soon as the heat ray touches an object, its surface is heated to an extremely high temperature, generating a flash of light and intense combustion. Note that in Wells's original description the heat ray does not disintegrate or impact on its target, effects that were introduced by Steven Spielberg in his 2005 film of the story.

ARCHIMEDES'S BURNING MIRROR

An intriguing aspect of Wells's description is that it sounds very similar to the burning mirror of Archimedes, the *locus classicus* of the death ray. Accounts of this might arguably constitute some of the earliest known science fiction, since it is generally doubted that such a device could really have existed. Archimedes was a third-century-BCE mathematician and inventor from the Greek colony of Syracuse in Sicily. Ancient accounts tell how his ingenuity in contriving defensive weapons held off a Roman invasion in 213 BCE, with devices such as a ship-wrecking claw based upon the principle of levers. Much later accounts – written a thousand years after the events they claim to describe – add a new and fantastical element to the tale. According to the twelfth-century CE Byzantine author John Zonaras, who claimed

to be paraphrasing the third-century-CE Roman author Dio Cassius, Archimedes's deadliest feat came when, "in an incredible manner, he burned up the whole Roman fleet". He accomplished this by "tilting a kind of mirror toward the sun [and thus] he concentrated the sun's beam upon it … [this] ignited the air … and kindled a great flame, the whole of which he directed upon the ships that lay at anchor in the path of the fire, until he consumed them all."

Another Byzantine scholar from around the same time, John Tzetzes, gives more technical details, describing how "the old man constructed a kind of hexagonal mirror, and at an interval proportionate to the size of the mirror, he set similar small mirrors with four edges, moving by links and by a kind of hinge, and made the glass the centre of the sun's beams". The power of these beams reduced far-off ships to ashes.

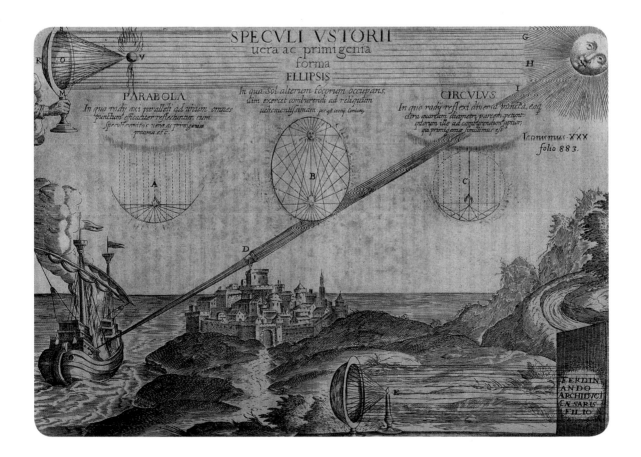

The burning mirror thus described sounds like a type of heliostat: a mirror that reflects the sun's rays onto a fixed point. The ancient Greeks knew that a parabolic mirror could reflect the rays of the Sun onto a single focal point, so a parabolic heliostat can concentrate solar energy to produce extreme heating and combustion. Using such a device as a weapon requires that the focal point can be controlled by the operator, which in turn would require some sort of adaptive optics. Indeed, this is what Tzetzes seems to be describing in his passage. So it seemed plausible that Archimedes's burning mirror could have existed. In fact, the thirteenth-century Franciscan monk Roger Bacon was sufficiently alarmed by the possibility that such a technology might have fallen into the hands of the Muslim opponents of the Crusaders that he wrote to Pope Clement IV to warn him of the danger that they "will use these mirrors to burn up cities, camps and weapons".

ABOVE The Comte de Buffon recreates the effect of the burning mirror, although in this illustration he is using lenses to focus sunlight.

OPPOSITE Ray guns quickly became one of the defining features of science fiction in the pulp era of the early and mid-twentieth century.

Later scholars both disputed and supported the possible reality of the burning mirror. In the seventeenth century Descartes dismissed the story as fantasy, but in 1747 George-Louis LeClerc, Comte de Buffon, claimed to have ignited a pine plank from 45 metres (150 feet), using an array of 128 mirrors. Experiments in Greece in 1975 and Germany in 2002 claimed to have proven the principle simply by getting a number of participants to hold up mirrors and work together to focus them on a target. An amusing variant of the proposition features in the 1958 Arthur C. Clarke story "A Slight Case of Sunstroke", in which 50,000 soccer fans wielding aluminium foil-covered programmes coordinate to focus the sun's rays onto a referee who makes an unpopular decision.

Such demonstrations notwithstanding, a parabolic heliostat would probably struggle to damage anything other than stationary, highly flammable targets at close range. To achieve the kind of directed-energy transfer across significant distance hinted at in the tale of the burning mirror we are lead once again to laser technology (see below).

TELEFORCE SUPERGUNS

Prior to the advent of laser technology, other types of energy weapon were explored by those operating in the borderlands between science and fiction. The dawn of the twentieth century was accompanied by an exciting proliferation of new types of ray (most notably the X-ray – see Chapter 12), along with strange new physics of the sub-atomic realm and extraordinary advances in the field of electromagnetism, such as radio and high-

voltage phenomena. This provided fertile ground for the public imagination and those that fed it, whether in the guise of fiction or sensationalistic journalism. A particular favourite of the interwar years was speculation about death rays, which would dramatically advance the killing power that had been unleashed during the First World War. Two of the great proponents of the death ray were scientists who had at least one foot in the hyperbolic world of speculative fiction: Nikola Tesla and Harry Grindell Matthews.

Tesla was a Serbian electrical engineer who had moved to America and helped create the fundamental technological architecture of the power age. He was one of the greatest inventors of all time, developing electrical motors, generators and transformers before moving on to more esoteric fields such as very high-voltage electrical phenomena. Although for various reasons he was not credited with their invention, Tesla was probably the first to demonstrate both X-rays and radio transmission. His public persona, burnished through dramatic demonstrations at expositions and world fairs, was of a kind of electrical wizard, able to toy with immense energies and strange forces. However, Tesla was also an eccentric whose visionary

The Return of the Cosmic Gun

By Morrison F. Colladay

(Illustrated by Marchioni)

There was a flash of violet light and the river seemed to explode with

fervour for increasingly ambitious but difficult-to-realize projects saw him become marginalized and gradually impoverished. This was the context in which he would make pronouncements and claims about potential death rays, which were propagated through the press.

A vacuum tube is a vessel from which most of the air has been expelled, leaving a partial vacuum occupied by only a very few gas molecules that become easily ionized by the application of electrical current. Invented in the Victorian era, it had provided the platform for discovery and demonstration of some of the earliest ray phenomena, such as X-rays and cathode rays (streams of electrons emitted from a cathode, the negative terminal of an electrical circuit). It was probably his work with vacuum tubes that prompted Tesla to develop the concept of a "teleforce ray": a beam or ray of some sort that could disable and destroy over great distances. Inevitably the media dubbed this the "death ray". As he

got older, and increasingly desperate to attract military sponsorship, Tesla's claims for the death ray became bolder and more outlandish. In 1924, Tesla proposed that his teleforce ray was capable of stopping aeroplanes in mid-flight. By 1934 he was boasting that his weapon could destroy 10,000 aeroplanes at a distance of 400 kilometres (250 miles), and that 12 teleforce stations positioned at strategic points around the country would provide complete protection for the United States. Tesla even claimed that the ray could penetrate into space and project a spot onto the Moon.

Here it is easy to spot forerunners of latter-day technology, such as the Strategic Defence Initiative of the 1980s, more popularly known as the "Star Wars" missile defence system. Its successor, today's "Son of Star Wars" missile shield programme, must also be included. Furthermore, the invention of the laser — an energy ray that has indeed been used to project a spot onto the Moon (in 1962) — bears an even closer resemblance to Tesla's claims. But does this mean that Tesla really did invent a death ray, and if so, was it a laser or something else?

In practice, Tesla seems to have anticipated several forms of ray or beam technology. His work with vacuum tubes, which included particle acceleration and X-ray production, could have suggested beams of X-ray or microwave radiation, lasers or particle beam weapons (where very highly accelerated ions, atoms or small groups of atoms are blasted at a target). One of Tesla's most authoritative biographers, Margaret Cheney, argues that "there is, in fact, no good evidence to suggest that Tesla anticipated lasers" but other writers

disagree, pointing out that a ruby laser (see below) is created in almost exactly the same way that Tesla's single electrode vacuum tubes worked. Alternatively, Tesla's research into the atmospheric effects of very high-voltage electrical fields may have suggested to him the possibility of using ionizing radiation or electrical fields to create channels of ionized air, which could perhaps act as electrical conduits for bolts of electrical energy.

Tesla may well have considered several of these options, but after his death the discovery among his papers of a 1937 treatise entitled *The New Art of Projecting Concentrated Non-Dispersive Energy through the Natural Media* confirmed that he had settled on a high-energy particle beam weapon of hugely ambitious design. In convincing technical detail, the paper describes a super-weapon consisting of a 30-metre (100-foot) high tower, within which was to be what is essentially a massive Van de Graaf generator, composed of a 67-metre (220-foot) long circular vacuum chamber to act as an airstream belt for generating electrostatic

charges of immense voltage. This, in turn, drove a supergun installed in a swivelling mount atop the tower, in which an open-ended vacuum tube, "sealed" at the open end by a high-velocity gas jet, would fire infinitesimally small particles of tungsten at a speed of 121,920 metres (400,000 feet) per second. Tesla had hinted at this "supergun" design in 1934 when he told a reporter, "This new beam of mine consists of minute bullets moving at terrific speed ... The whole plant is just a gun, but one which is incomparably superior to [those of] the present." Some elements of this scheme are mirrored in modern mobile phone microwave transmitters, but it is not clear whether Tesla's machine would have worked, or how it would have overcome the problem of scattering and dispersion in the atmosphere.

In fact, the most likely scenario is that Tesla's claims about superguns and teleforce rays were little more than science fiction, concocted to boost his flagging public profile and possibly to attract investment. This was even more likely the case with the claims of the British inventor

Harry Grindell Matthews, a kind of cut-price Tesla who became closely associated with death ray flim-flam.

THE DEATH RAY MAN

Grindell Matthews was a British radio pioneer whose career initially promised great things, and who could be remembered today as a huge success if fate had only been kinder. In the early 1920s he invented a highly sophisticated system for bringing sound to the movies, which, if backed by investors, might have made Britain the centre of the global film industry. However, Grindell Matthews's venture into talking pictures came to nothing, and in 1923 he launched a bold new bid to gain public attention and financial backing by touting a potential new military technology.

The newspapers at the time were full of reports of how French aeroplanes were mysteriously failing in mid-flight over German soil. Grindell Matthews believed he knew why, after noting that the aircraft had been forced down while passing close to powerful German radio transmitters. It was the emanations from these transmitters that had caused the aircraft's engines to stop, and Grindell Matthews believed that he could harness this effect to create a new and devastating weapon, one that would safeguard Britain against the growing menace of aerial bombardment and, eventually, render war effectively impossible.

Within weeks Grindell Matthews claimed to have developed an apparatus that could explode gunpowder, light a lamp and kill vermin from a distance of 20 metres (65 feet). As his experiments progressed he claimed that he had melted plate glass, stopped motorcycle engines and even accidentally knocked unconscious an assistant who strayed into the path of the beam. Reporters soon got wind of the new weapon and, when he obligingly staged a demonstration for one of them, a media firestorm was unleashed. Grindell Matthews was dubbed "Death Ray Matthews", causing him to lament melodramatically, "Why do the Press call me the Death Ray Man? Am I a monster of destruction, seeking only to turn what brains have been given me for the annihilation of others?"

In fact, he willingly manipulated the media to gain publicity and pressurize the authorities into funding him. Sensationalist reports were fed by Grindell Matthews's increasingly ambitious claims. "I am confident," he declared, "that if I have facilities for developing it I can stop aeroplanes in flight – indeed I believe the ray is sufficiently powerful to destroy the air [and] explode powder magazines."

Hectored by the press and fearful that a foreign power might purchase Grindell Matthews's death ray, the British Air Ministry reluctantly agreed to a demonstration. In April 1924 Grindell Matthews staged one at his own laboratory, but the military observers were unimpressed by what they saw, which amounted to little more than lighting a bulb and stopping a motorcycle engine.

M. MATTHEWS
laisse photographier son rayon ardent

ON A ENCORE PRÉSENT A LA MÉMOIRE LE FAMEUX RAYON INVISIBLE SI DISCUTÉ, AU MOYEN DUQUEL ON POURRAIT, DE TERRE, BRULER DES AVIONS VOLANT EN PLEIN CIEL OU FAIRE EXPLOSER A DISTANCE LES DÉPÔTS DE MUNITIONS. VOICI M. GRINDELL MATTHEWS, SON INVENTEUR, DANS SON LABORATOIRE.

Grindell Matthews became evasive and argumentative when changes to the test situation were suggested, and in a meeting at the Air Ministry the next day officials agreed that something smelt rotten. Testimony from a man named Appleton, who may have been an MI5 agent, effectively accused Grindell Matthews of being a con man who was "working the press, but had now lost control of it". When further tests were proposed to Grindell Matthews, he insisted that none were necessary.

Grindell Matthews's side of the argument was that he simply did not trust the government to stage a fair and impartial test. He even claimed that unknown sources had warned him that the tests were a trap. The whole affair descended into farce on 27 May, when an Air

OPPOSITE Grindell Matthews in full "mad scientist" pose, photographed as part of a promotional piece about his "*rayon ardent*" (scorching ray).

BELOW Further photographs from the publicity material showing an earlier Grindell Matthews' invention (a kind of spotlight) and supposed pictures of his ray in action.

L'EXPLOSION DANS LE LABORATOIRE

Notre scène de droite représente l'explosion de poudre à distance dans le laboratoire. Si intéressant que soit le film, il ne saurait constituer une démonstration scientifique.

LE RAYON A LA POURSUITE DE L'AVION

M. G. Matthews a laissé filmer ses appareils, dont nous avons donné il y a quelques mois un fort exact dessin. Nous publions aujourd'hui une épreuve du film où l'on voit (à gauche) la boîte mystérieuse d'où sortirait le rayon et les groupes électrogènes qui contribueraient à sa production. Ajoutons que l'expert officiel américain, envoyé par la Chambre des Représentants, vient de déclarer, comme les experts du gouvernement anglais, que le rayon n'existait pas.

202

203

Ministry official arrived at Grindell Matthews's lab to make a last-ditch attempt to arrange a demonstration. However, he found only a group of angry businessmen – Grindell Matthews's backers, frustrated by the lack of return on their investment – who had that very morning won a court injunction assigning them all rights to the "death ray". All concerned piled into a car and followed Grindell Matthews's trail to Croydon aerodrome, missing him by seconds as he boarded a 'plane to France.

The controversy continued with the Press supporting Grindell Matthews and the government insisting that he must submit to proper testing. Other investors – French and British – offered large sums of money to the inventor, but despite his claims for the device he consistently refused to stage demonstrations to anyone's satisfaction. Interest in the United Kingdom eventually petered out but Grindell Matthews found the Americans more receptive, agreeing to star in a docudrama called *The Death Ray*. To both the American and British media he continued to make grandiose claims, explaining that he hoped to extend the range of the beam to 8–13 kilometres (5–8 miles), and that "so soon as my ray touches a plane it bursts into flame and crashes to the earth".

One of the many unanswered questions regarding the death ray was its mechanism, but several hints emerged. At one point Grindell Matthews explained that it used two beams of light, one a "carrier" and the other as a "destructive current". He also said that it was "able to convey electrical power through space as is done by the thunderstorm". Such clues suggest that Grindell Matthews was attempting to use either light (for example, a beam of UV) or very high voltages to create a channel of ionized air, which could then function as a conductor through which an electric charge could travel to and disable the target. This is essentially what happens in a thunderstorm: massive voltages between the thunder cloud and the ground cause "streamers" of plasma (ionized air) to develop, and electric current travels along these, creating a lightning strike.

Passing current along an ionized channel had been one of the mechanisms considered by Tesla in devising

his own death ray, the teleforce ray. While Grindell Matthews explicitly acknowledged this debt, admitting in 1924 that "the present experiments [were] inspired by Tesla", he also asserted that his ray "differs considerably from [Tesla's] plan". More generally he called Tesla "the man who has been my inspiration and my greatest source of hope – [he is] to me the greatest man on earth today". Tesla did not return the compliment, commenting in 1934, after hearing details of Grindell Matthews's death ray, "It is impossible to develop such a ray. I worked on the idea for many years before my ignorance was dispelled..." In this, Tesla was almost certainly correct. If Grindell Matthews's plan had been to create a plasma channel and send electricity along it, he would have found it effectively impossible – it is certainly beyond the most powerful modern technology to emulate a lightning strike, let alone guide it accurately to hit a distant and moving target. Like Tesla's claims, those of Grindell Matthews belong more to speculative fiction than fact, and indeed death rays and similar energy weapons were well on the way to becoming a staple of science fiction.

THE INVENTION OF THE LASER

As early as the 1930s, directed-energy weapons had already become a cliché in science fiction. John W. Campbell's 1932 story "Space Rays" involved so many beam weapons that magazine editor Hugo Gernsback was driven to comment, in an introductory rubric, "If he has left out any coloured rays, or any magical rays that could not immediately perform certain miraculous wonders, we are not aware of this shortcoming in his story." Gernsback chose to interpret Campbell's promiscuity with rays as a sort of satire, "an earnest

OPPOSITE Poster for the classic 1951 film *The Day the Earth Stood Still*, in which an alien robot can unleash a devastating energy beam.

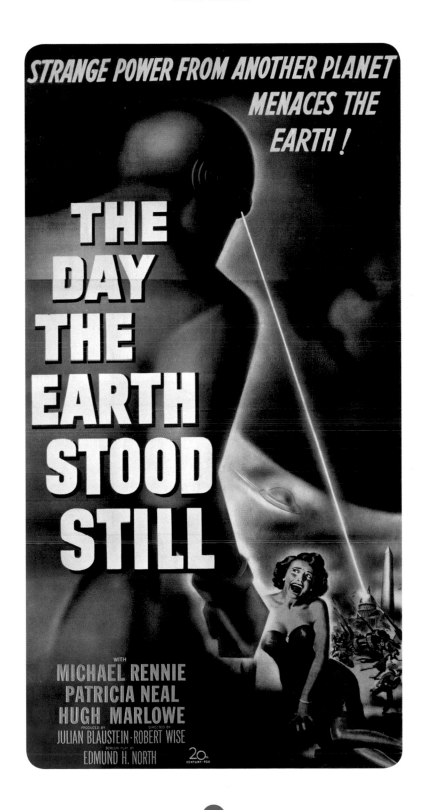

way to burlesque some of our rash authors to whom plausibility and possible science mean nothing", a particular bugbear of many "serious" sci-fi writers at the time. Gernsback disparaged these "impossible rays" as "a preposterous gimmick", complaining that "hurling fifteen million horsepower from one space flyer to another means nothing".

It would take "possible science" nearly 30 years to approach, even as a demonstration of principle, what is today considered the classic directed-energy weapon, with the first laser not invented until 1960. The term laser did not even exist until 1959, when it was coined as an acronym for "Light Amplification by Stimulated Emission of Radiation". A laser is a beam of light that is all of the same wavelength and all moving in the exact same direction (it is said to be coherent). This allows it to deliver energy to a target with pinpoint accuracy. The keys to generating a laser are getting a material – known as the lasing medium – to pump out photons of light that are all of the same wavelength, and then getting them all to move in the same direction. The first laser, created by Theodore Maiman at the Hughes Research Lab in Malibu, California in 1960, used a rod of synthetic ruby as a lasing medium, around which was wrapped a coil of xenon flash lamp (essentially a sophisticated fluorescent light tube). The flash lamp pumped energetic photons into the ruby, which excited the electrons around the atoms of the ruby so that they in turn emitted energetic photons of a single wavelength. The ruby had mirrors at each end, so that the photons bounced back and forth along the rod, cohering into a single beam. At one end the mirror was only half-silvered, so that a powerful enough beam could pass through as a laser. This laser was created on principles demonstrated a few years earlier with microwaves (giving a maser), and it was Charles Townes, creator of the maser, who would be one of those jointly awarded the first patent for the laser.

Lasers inspired tremendous excitement in both the science fiction and science fact communities, but the former would quickly outpace the latter. Lasers replaced ray guns as the default science fiction weapon, a status

OPPOSITE TOP Stormtroopers firing laser weapons in a still from the 2015 movie *Star Wars: The Force Awakens.*

OPPOSITE BOTTOM An early laser experiment, with a beam of coherent light emitted from a rod of lasing medium.

ABOVE Targeting lasers are now a routine part of gun-sighting equipment.

ABOVE Concept art showing Lockheed Martin's truck-mounted ATHENA
battlefield laser, technology that is at or near operational deployment.

they would maintain until the 1980s, when films like *Blade Runner* (1982) and *Aliens* (1986) featured slug-spewing military hardware in tune with a grittier depiction of future tech. However, prior to this development, audiences became accustomed to iconic weapons such as the phaser pistols of *Star Trek* and Han Solo's "Heavy Blaster" laser sidearm.

DAZZLERS AND DRONE DEFENCE

In the real world, millions of dollars were pumped into laser research, particularly in the hunt for military applications. Laser weapons offer game-changing potential: speed-of-light velocity; pinpoint accuracy; straight-line trajectory; adaptive profile to match different target types, materials and intended damage; low cost per shot; and inexhaustible ammunition. Military research sought to develop lasers for blinding and dispersing human targets; cutting through armour; disabling sensors and targeting devices, and intercepting even the fastest targets, including intercontinental ballistic missiles. This last application led to the US military's "Star Wars" project of the Reagan era, which envisaged space-based lasers disabling or destroying Soviet missiles to create an impenetrable missile shield over America and her allies.

The practical hurdles to effective laser deployment have meant that almost all of these ambitions have remained unfulfilled. Balancing often competing demands – such as power and operating temperature, range and accuracy – has proved to be incredibly difficult. Over long distances, for instance, atmospheric fluctuations – including those caused by the laser itself – disrupt the coherence of the beam and cause it to lose focus. A continuous laser beam can also diminish its own impact on a target by creating a cloud of plasma that acts as a shield, so that armour-piercing applications often use a pulsed approach. Other considerations include the nature of the lasing medium – solid media for "solid-state lasers" are easier to work with but prone to heating, but chemical lasers often use dangerous and volatile reactants, cooling techniques, target acquisition and tracking, and, above all, power supply.

Militarily effective lasers require very high energies, and this, in turn, limits their portability and potential platforms, since large battery packs and/or power plants are needed. Simple physics thus make it unlikely that the laser sidearms of science fiction will ever become reality, because even the most energy-dense power packs would be both incredibly bulky and extremely dangerous, since if hit during a fire fight they would be likely to produce a colossal explosion. A common criticism of such a system is that it would be more effective to use such power packs as grenades than to power light weapons. An eccentric near-exception to this rule was a Soviet programme to develop flashbulb-powered single-pulse laser pistols to be issued to cosmonauts, but it never got far.

Apart from "dazzle" applications, in which lasers are used to blind targeting technology – for example, as a defence against anti-ship missiles – lasers still have not made it onto the battlefield. In the next year or so this situation may, finally, change. The US military has programmes such as the Navy's High Energy Laser and Integrated Optical-dazzler with Surveillance programme, or HELIOS, which is intended to deploy both dazzling and offensive capabilities against unmanned aerial vehicles (UAVs) and small boats, and is supposed to be fitted to an active duty vessel in 2020. HELIOS itself is based on laser platforms such as the Advanced Test High Energy Asset (ATHENA) prototype which, in a celebrated 2015 test, was able to disable a truck from a mile away. Knowing for sure whether such weapons are currently operationally deployed or have been used in warfare is complicated by military secrecy, but even if lasers do become part of the standard military arsenal over the next decade, it will be as part of anti-UAV and other drone defence systems, rather than slicing through tanks or aeroplanes, as in the public imagination.

Another obstacle to battlefield deployment of lasers is the international convention on weapons. Weapons that cause drawn-out suffering are considered inhumane, and since laser technology is not currently powerful enough to cause instant death of a human target, laser weapons are banned from lethal anti-personnel use. But these self-

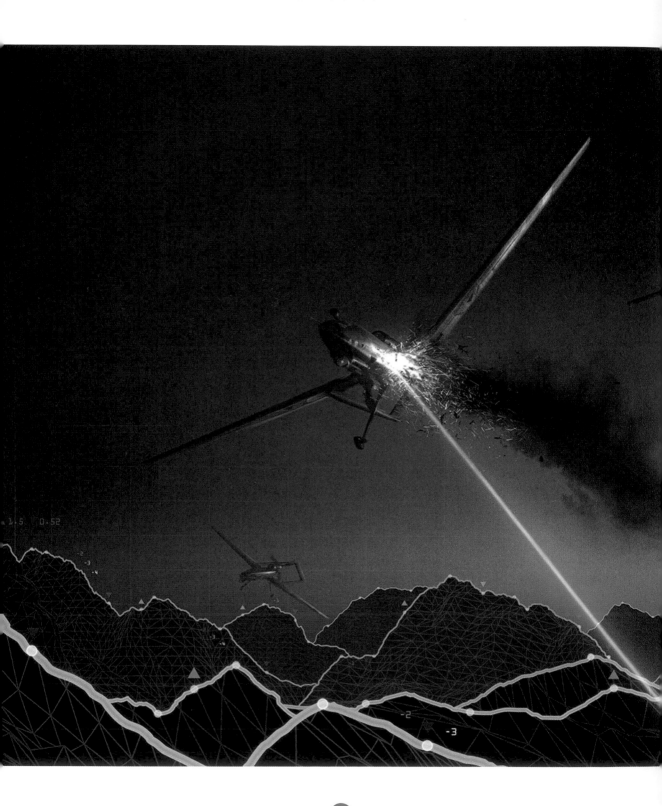

same lethality issues have made lasers a technology of interest for those researching "active denial" solutions: a euphemism for non-lethal anti-personnel weapons that can disperse crowds and drive people away. Lasers, including, potentially, microwave lasers, are among various directed-energy weapons (including sound weapons) that cause pain and distress to which a target responds by moving away. Such crowd-control lasers are thus similar in mechanism to the heat rays deployed by H. G. Wells's Martians, causing heating and burning on contact. The difference is only one of power rating, with the Martians having apparently overcome many technical hurdles such as power supply, efficiency of energy conversion, overheating/cooling and adaptive optics, not to mention ethics.

LEFT Conceptualization of the most likely near-future reality of military lasers disabling UAV drones.

04

DRONES AND AUTONOMOUS WEAPONS:
FROM TESLA'S TELAUTOMATONS TO THE PREDATOR

The Predator drone is now among the most ubiquitous symbols of the American military, and drones — unmanned, remote-controlled vehicles — are rapidly becoming major components of advanced military forces around the world.

Although drones include surface and sub-surface ships and ground vehicles, the most familiar and common type of drone is the unmanned aerial vehicle (UAV): in 2005 only five per cent of US military aircraft were drones, whereas today UAVs almost certainly outnumber manned aircraft in the American military.

An argument could be made that the first unmanned aerial weapons were the bomb-bearing pilotless balloons launched by the Austrians in 1849 during an assault on Venice. However, these clearly do not match the concept of drones as generally understood — controlled but unmanned vehicles. The First World War engendered attempts to develop remote-control vehicles, including remote-controlled anti-Zeppelin aircraft designed, but never successfully developed, by British engineer Archibald Montgomery Low. Another example is American inventor Elmer Sperry's 1915 aerial torpedo, a lightweight airframe packed with explosives and stabilized with a gyroscopic guidance mechanism. However, most historians of weapons technology trace the advent of the drone back to the Second World War, when aircraft such as the German Argus As 292, the British Queen Bee and the American TDR were fitted with radio controls to allow remote piloting.

In spite of all this, the conceptual roots of drones and autonomous weapons as understood today lie in the borderland between science fiction and science fact — a territory explored almost without parallel by the Luxembourgish-American magazine editor, science-fiction fan, author and science popularizer Hugo Gernsback, who called it "scientifiction". His magazine *Electrical Experimenter* was a prominent exponent of this genre, mingling articles on cutting-edge technology with speculations that were supposedly grounded in reality, but which clearly ventured into sci-fi.

THE AUTOMATIC SOLDIER
Gernsback dreamed up at least two iterations of autonomous or drone-like weapons, most notably the "Automatic Soldier" that he outlined in the October 1918 issue of *Electrical Experimenter*. Writing in the dying days of the First World War, Gernsback pointed out that "as science advances, and as all sorts of infernal machines are thrown into a modern war, the men in the front-line trenches become less and less anxious to bear the full brunt of high-explosive shells, gas attacks, liquid fire and what not". What was required, he suggested, would be "some sort of a soldier who was bomb- and shell-proof and who did not mind either liquid fire or the most vicious kind of gas".

Citing a patent sought by an unnamed "Danish engineer", Gernsback went on to describe what he called an "automatic soldier" — an automated, armoured, stationary weapons platform. Looking, in the

OPPOSITE "The Flying Buzzsaw", a dramatic, if impractical, early vision of an aerial drone in action, in a Gernsback-edited pulp magazine.

AIR WONDER STORIES

APRIL
1930
25 CENTS
Canada 30c

HUGO GERNSBACK
Editor

A GERNSBACK PUBLICATION

"The Flying Buzz-Saw"
by H. McKay

Other Science Aviation Stories by Edmond HAMILTON
George Allan ENGLAND   Lowell Howard MORROW

OCT.
1918
15 CTS.

ELECTRICAL EXPERIMENTER
SCIENCE AND INVENTION

OVER
175
ILLUST.

REG. U.S. PAT. OFF.

**THE AUTOMATIC
SOLDIER**
SEE PAGE 372

accompanying illustration, remarkably like a Dalek (an alien-controlled robotic warrior from the TV series *Doctor Who*), the automatic soldier was to be constructed from a double layer of "shell-proof Tungsten steel", comprising a retractable dome securely embedded within a front-line trench. It was to be operated by wireless remote control, by controllers situated at some distance back from the front line, guided by aerial observers. Once triggered, the dome would rise up above the lip of the trench and unleash its firepower on unfortunate enemy combatants approaching the trench. Gernsback suggested that an array of automatic soldiers would be deployed across the frontline, most equipped with machine guns, but every sixth one with a poison-gas gun. He envisaged the devices running on batteries and using magnetic and compressed air actuators.

Gernsback confidently asserted that "we would be very much surprised if the automatics would not make their appearance soon at strategical points along the

ABOVE Ronny Cox, as sinister OmniCorp executive Dick Jones, with "Enforcement Droid" ED-209 in the 1987 movie *RoboCop*.

OPPOSITE Gernsback's Automatic Soldier – essentially an automated machine-gun post – in action.

line". In practice, automatic gun emplacements never became part of the standard battlefield deployment, partly because mobile armour would soon arrive to move the tactical status quo away from trench warfare. In some settings, however, automatic machine-gun turrets are now commonplace: for instance, on naval ships or as part of anti-aircraft batteries. The Automatic Soldier also calls to mind a notable latter-day sci-fi interpretation: the automatic machine gun emplacements of James

Cameron's *Aliens* (or at least its "Director's Cut").

The Automatic Soldier was not Gernsback's only vision of a future weaponized drone. In the May 1924 issue of his magazine *Science and Invention* he proposes a remote-control, unmanned police robot, which he calls a Radio Police Automaton (the term "robot" was only just beginning to come into general usage). It was very much a conceptual predecessor of the "Enforcement Droid" robots of Paul Verhoeven's 1987 film *Robocop*. The accompanying illustration depicts a somewhat dystopian vision of terrified crowds of humans being menaced by giant robots brandishing whirling billy clubs for hands. A cutaway shows that the robot was to move on caterpillar-tracked feet, be powered by petrol engines, controlled by radio remote control and equipped with loudspeakers, powerful lights and tanks of tear gas.

TESLA'S TELAUTOMATONS

Gernsback acknowledged that one of his primary sources of inspiration was the Serbian-American inventor Nikola Tesla, and Gernsback's magazine provided a platform for Tesla to publish speculative articles and even parts of his autobiography. In fact, Tesla had not only previously anticipated some of the concepts of Gernsback's autonomous weapons – he had built working prototypes of them fully 20 years before, when he presented his Telautomatons to the 1898 Electrical Exposition in New York City's Madison Square Garden.

Tesla was a profligate genius who dreamed up many futuristic inventions, such as a system for blasting message containers around the world through water-filled tubes, a giant space hoop circling the Earth at the Equator, and a particle accelerator death ray (see page 40). His pioneering work on high-voltage electrical fields generated an assortment of strange phenomena, including X-rays, wireless transmission of power, and radio waves. As early as 1893 Tesla had invented equipment that could transmit and receive radio waves, and gave what is now recognized as the first demonstration of radio transmission. This was two

ABOVE Serbian-American inventor and futurist Nikola Tesla in his prime.

OPPOSITE Interior view of one of Tesla's radio-controlled boats or telautomatons.

years before Marconi's first efforts in the field, but Tesla allowed the less perfectionist Marconi to steal a march on him in the production of a commercially available system. In fact, to achieve his 'breakthroughs' Marconi relied on Tesla's patents and theories, but he never admitted it and actually challenged Tesla's patents in an epic court battle that was eventually decided in the Serbian's favour, though not until both men were dead.

His experiments with radio had convinced Tesla that it would be possible to wirelessly transmit control signals to a distant device, and by 1898 he had evolved this work into a complete concept of telerobotics, presenting a remotely controlled model boat that he called a "telautomaton". This prototype was much more than a simple toy. Not only did it run off wirelessly transmitted power, but it also had a sophisticated control system of lights and motors so that it could perform complex operations, but only in response to correctly coded broadcasts on multiple frequencies, making it the forerunner of any device that uses scrambled or protected radio signals, from cable television to garage door openers, not to mention its direct descendants, today's UAVs.

The 1898 prototype was just the beginning. Tesla dreamed of a much more advanced robot with significantly greater autonomy – in other words, some form of artificial intelligence (AI). "The time is not distant," he boasted in 1919, "when I shall show an automaton which, left to itself, will act as though possessed by reason – it will mark the beginnings of a new epoch in mechanics." As far as Tesla was concerned, his telautomaton was the first non-biological life form on the planet, and he had created artificial life (AL). He pictured fleets of his robot boats and, in later life, aeroplanes, which would receive power and instructions from distant control stations but would also know where they were and act autonomously. Human warriors would not be needed, and war would be fought entirely by machines.

LINE OF SIGHT

Unfortunately, Tesla's 1898 presentation of telautomatons was not a success. The people who saw his demonstration at the Electrical Exposition in Madison Square Garden, in which he made the robot boat move around a tank of water and respond to commands and questions from the audience, failed to grasp the magnitude of his achievements. Not only were they witnessing one of the first practical demonstrations of radio, they were also being introduced to robots, remote control, remote transmission of power and concepts of AI. At the same exposition Marconi was winning plaudits for a comparatively crude demonstration of short-range

ABOVE The Queen Bee, an early pilotless aircraft, shown in 1935, with the remote control station in the foreground.

OPPOSITE A German Goliath — an explosive-carrying, remote-control drone — from the Second World War, nicknamed the "beetle tank" by Allied troops.

radio, yet Tesla's far more advanced achievements went overlooked. He was too far ahead of his time, but he was also overreaching himself.

For all his claims about AI and AL, Tesla's telautomatons were nowhere near being autonomous, and AI of the kind he was describing remains beyond the grasp of technology. Since there was no kind of video technology available at the time, without the ability to direct themselves, the telautomatons relied entirely on remote control by an operator working within visual range.

Although he overreached and overpromised, Tesla's pioneering work anticipated many of the suite of technologies that lie behind today's drones. Work to create remote-control military vehicles continued, with the efforts of the British inventor known as "Death Ray" Grindell Matthews (*see* page 42) winning brief notoriety.

In the early years of the First World War, Grindell Matthews experimented with remote control using selenium, a photo electric metal that becomes conductive when illuminated. He fitted a selenium-activated remote control to a boat he christened *Dawn*,

so that it could be remotely controlled by a beam of light. Equipped with its "selenium pilot", this small boat could be steered and operated with the beam of a searchlight, and he eventually got it to work at ranges of up to 3,000 metres (3,000 yards) in "diffused daylight" and 8 kilometres (5 miles) at night.

With the First World War raging, the Admiralty put out a call for new technologies of exactly this nature. Grindell Matthews put on a successful demonstration for the military and in 1915 they bought the technology from him for the princely sum of £25,000. Grindell Matthews had big ideas for his light-controlled technology, formulating plans for remote detonation of explosives and remote-controlled mini-airship torpedoes, which could be directed against enemy zeppelins. However, his speculations went nowhere and the Admiralty never developed the Dawn concept.

Leading up to, and during, the Second World War, speculative incarnations of the drone – like those of Gernsback, Sperry and Grindell Matthews – gradually gave way to operational ones. Radio remote-control

toy aircraft enthusiast (and British film actor) Reginald Derry produced thousands of remote-control aircraft for the US military, which were used in training anti-aircraft gunners. The success of his OQ2 radioplane even managed to launch the career of Marilyn Monroe, who was first talent-spotted when a photographer went to document production of the OQ2, spied Norma Jeane Dougherty working at the Denny factory, and convinced her to pose with a drone propeller. In Europe, the Germans developed a low-profile, remotely-operated tracked vehicle named the "Goliath Light Charge Carrier", designed to be a mobile mine that would carry explosive loads under tanks or into buildings and then detonate. Allied troops nicknamed these machines "beetle tanks". Interestingly, they were hamstrung by many of the same drawbacks that still hamper modern military robots, being too vulnerable and unreliable to justify the considerable expense of the hardware.

OPPOSITE TOP A USAF "Lightning Bug" AQM-34L reconnaissance drone in operation over North Vietnam in the 1960s.

OPPOSITE BOTTOM The "Sea Hunter", a US military drone, gets underway in 2016; the people are only on board because of a christening ceremony.

ABOVE The Reaper drone, one of the most widely deployed and successful UAVs, is a Hunter-Killer.

HUNTER-KILLERS

Today's most familiar drones – the Predator- and Reaper-style UAVs deployed by the American and many other militaries – descend from glider-stye drones developed by the Israelis in the 1970s and 80s. It is these vehicles that most closely correspond to the speculative visions

of pioneers such as Tesla and Gernsback; they are relatively cheap, very common and highly effective, and they are changing the nature of warfare, prosecuting conflicts while their operators sit in bases on distant continents, commuting between home and workplace like ordinary office workers. Compare today's reality with the future foretold by a 1981 children's book, *World of Tomorrow: Future War and Weapons*, which predicted that future "robot battalions" would be directed by "leaders ... safely situated far behind the fighting", and foresaw with startling accuracy "small pilotless aircraft packed with sensors and detectors of all kinds to plot enemy positions, measure forces and eavesdrop on their communications".

However, it is science fiction that continues to dominate the public perception of drones – at least in terms of the expectations and anxieties surrounding them. In this context, the exemplar is the Hunter-Killer of James Cameron's terrifying vision of an AI-inflicted

ABOVE A Raven surveillance drone is launched by an American marine in Afghanistan in 2009.

OPPOSITE A heavily armed hunter-killer robot prowls for humans in a still from the 2003 movie *Terminator 3: Rise of the Machines*.

near-future apocalypse, in his *Terminator* films (primarily *The Terminator* (1984) and *Terminator 2: Judgment Day* (1991)). In this version of the future a global AI network called Skynet becomes self-aware and attempts to wipe out humanity to preserve itself. It launches nuclear weapons and builds and deploys a networked robot army to mop up survivors. The primary "soldiers" of this robot force are Hunter-Killer drones, autonomous military robots that include tank-based ground vehicles and heavily-armed fan-rotor aircraft. The Wachowski

siblings' *Matrix* films present another post-AI apocalypse vision, in which humans are hunted by airborne robots that resemble squid.

These visions of the future explore the looming ethical ramifications of truly autonomous weapons – those that make their own decisions about use of lethal force. Some of the functions of existing drones are already autonomous, but decisions about acquiring and, crucially, engaging with targets are still made by humans. However, many nations are developing systems intended to automate some or all of the steps in this process. For instance, the British Royal Air Force has a programme named Future Combat Air System, which explicitly states plans to develop the BAE Systems Taranis drone into one that is at least capable of being a lethal autonomous weapon. Intermediate steps on the road to fully autonomous "kill decisions" might be systems that can identify, locate and acquire targets, so that a human operator is only required to give a green light to launch a strike. Such systems are almost certainly already well developed. Concerns about such developments prompted the creation, in 2013, of the Campaign to Stop Killer Robots. The shadow of *Terminator*'s Hunter-Killers looms large over current discourse.

PART 2
LIFESTYLE & CONSUMER

05

CREDIT CARDS:
THE "CASHLESS SOCIETY" PREDICTED IN 1888

Money is part of the fabric of any society, and perhaps this is why science-fiction authors have often taken it for granted and failed to devote much attention or ingenuity to future finance.

The most common concession to futurity is to recast varying national currencies as universal credits, so that credits or "creds" have become a sci-fi cliché. There have been some interesting exceptions to this rule, in which sci-fi authors have asked what might happen in post-scarcity economies that result from matter duplication technology (see Chapter 7), or explored novel approaches to money and the medium of exchange. Examples include a system based on virtue, in which credits are awarded for good deeds, featured in Patrick Wilkins's 1954 story "Money is the Root of All Good"; a planet of immortals where the money system is based on the only commodity of ultimate value: a poison that can release the inhabitants from their mortal torment, in Neal Asher's 2002 novel *The Skinner*; and a system in which wealth is distributed based on personal prestige or social standing, as in Jack Vance's 1961 story "The Moon Moth", or Cory Doctorow's 2003 novel *Down and Out in the Magic Kingdom*.

PUNCH CARD PROSPERITY
One of the most prescient treatments of money in sci-fi is also one of the oldest, for in his 1888 novel *Looking Backward*, Victorian author Edward Bellamy predicted and even precisely named the credit card. This book caused a sensation at the time, although it was only one of a host of similarly premised novels in which a sleeper awakes in the future and provides a didactic

ABOVE The front cover of Bellamy's 1888 novel, showing his sleeper awakened in the year 2000, and feeling rather overwhelmed.

survey of the state of things to come. In Bellamy's book, a sleeper named Julian West awakes in the American city of Boston in the year 2000 to find himself amidst a socialist utopia, in which citizenship is defined by and derived from participation in joint labour. In this system, money has been abolished and replaced with credits, apportioned to every individual in equal measure "corresponding to his share of the annual product of the nation". Instead of money, each citizen is issued with a punch card that records how many credits that person has left, after "he procures at the public storehouses, found in every community, whatever he desires whenever he desires it".

"Perhaps you would like to see what our credit cards are like?" suggests Julian's guide, showing him "a piece of pasteboard" and explaining, "the value of what I procure on this card is checked off by the clerk, who pricks out of these tiers of squares the price of what I order". This decidedly lo-tech system, which might more properly be called a debit card, since it is used to spend assets already held, might seem from a cynical modern perspective to be an invitation to fraud and malign manipulation. But as Bellamy explains, since wealth and money have been eliminated, so have crime and social disorder.

Bellamy's credit cards are honoured anywhere in the world, much like those of today, and he even has a kind of early socialist Amazon delivering goods by pneumatic tube (the Victorian era's last word in bleeding-edge technology) from central warehouses. These tubes connect to government-run stores, but also to people's homes, and have wide enough bores to handle large items.

TALLY STICKS TO CHARGE PLATES

Does this make Bellamy the father of the credit card? The people behind an eccentric statue erected in the Russian city of Yekaterinburg, in 2011, evidently think so. Proudly claimed to be the world's first and only monument to the credit card, it is 2 metres (6 feet 6 inches) high and made of cast iron coated with bronze. The statue takes the form of a hand holding

ABOVE The idiosyncratic monument to the credit card in Yekaterinburg, with a card bearing the name "Edward Bellamy".

ABOVE English tally sticks, c.1440, were a method of accounting for loans and payments.

OPPOSITE A 1973 advert for what was then known as the Interbank Master Charge card, which later became MasterCard.

a credit card, upon which is printed the name Edward Bellamy. Created by sculptor Sergey Belyaev, it was commissioned by the local VUZ-Bank.

What is less clear is whether the actual creators of the credit card had Bellamy's tale in mind. Credit cards have ancient antecedents in the form of other types of media for transferring value that are not themselves of inherent value (and are thus different from coins), beginning with tally sticks. These are sticks marked or notched to keep a tally of a commodity (such as bushels of grain or units of currency). These can be broken in two to create a uniquely fingerprinted record of financial obligation, with half a tally stick being redeemable on presentation and matching with the other half. The impossible-to-forge jagged line marking the break between the halves is the primitive equivalent of the encryption that makes a credit card secure.

Credit cards as we know them are specific credit instruments, which allow a single card to be used to carry debts incurred through transactions with multiple retailers and different organizations. Their immediate forebears were "charge plates" – metal plates or cards specific to individual businesses, which meant

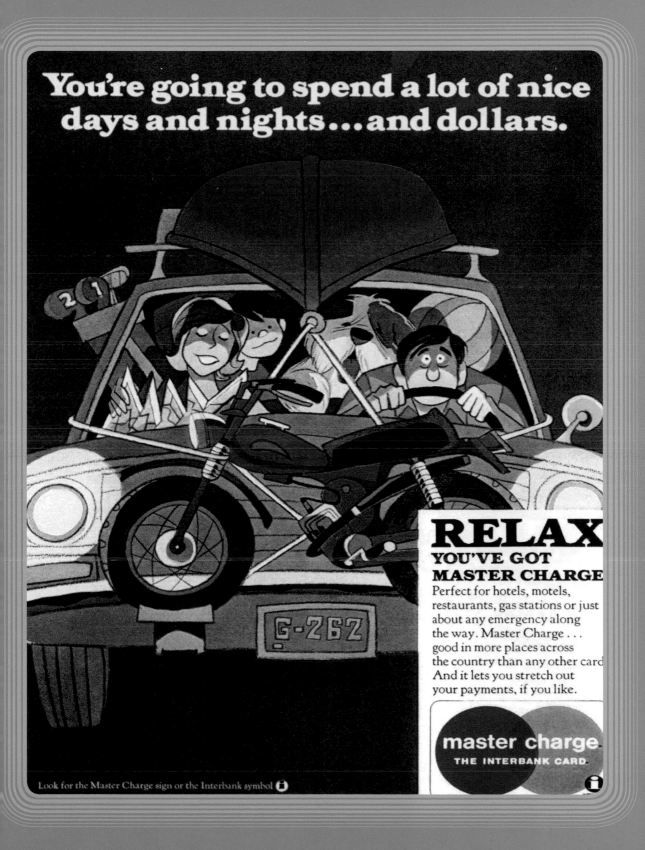

customers could make transactions without needing ready cash. Essentially these were markers of creditworthiness, aka financial trustworthiness – tokens to indicate that a merchant could trust this customer to be good for their debt. This harks back to the origins of commercial exchange, before the advent of money, which, contrary to popular myth, lay not in barter but in social networks of trust, promise, debt and obligation. At least one sci-fi author imagined a hi-tech future version of such a system; in Eric Frank Russell's 1951 story "… And Then There Were None", the currency is "obs" or "obligations".

In 1950 Ralph Schneider and Frank McNamara, founders of Diners Club International, amalgamated charge plates for multiple businesses into a single plate that could be used in many different restaurants; it was a charge card, rather than a credit card, since the debt had to be cleared rather than carried. In 1958 the Bank of America introduced the BankAmericard, the first successful and widely accepted credit card, and American Express launched their card in the same year, going plastic the following year.

SMART CARDS

Credit cards have continued to develop since Amex went plastic, and some of these developments have been foretold in science fiction. Smart cards (plastic credit and other types of card with integrated microprocessor chips) were invented in the 1970s, but until the mid-90s they were mostly restricted to telephone debit cards (cards that carried prepaid credits for making phone calls from public phones). In his 1988 novel *Mona Lisa Overdrive*, however, William Gibson makes smart cards ubiquitous – for example, one of his protagonists brandishes her "platinum MitsuBank chip" when asked to pay for something.

In the true cashless society now dawning, even the card part of a debit or credit card becomes redundant. Wirelessly communicating microchips are increasingly small, cheap and ubiquitous, so that cashless payment can now be performed with smartphones, or even with microchips implanted under the skin. Using your smartphone as a payment device was specifically foretold as early as 1966, in Frederik Pohl's book *The Age of the Pussyfoot*, although his version of the smartphone

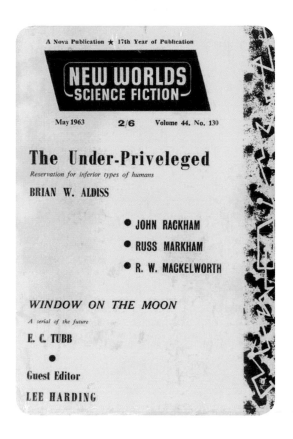

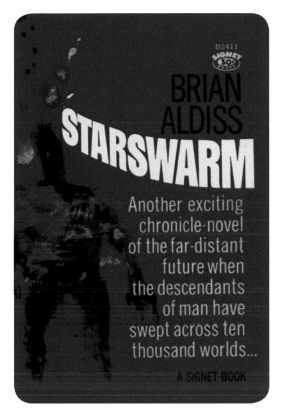

or personal digital assistant is called a "joymaker" (see page 203 for more on Pohl's remarkably prescient joymaker). "To use your joymaker as a credit card," reads a manual presented to the novel's protagonist, "you must know the institutional designation and account spectrum". This technobabble is presumably akin to knowing your PIN.

Another development trailed by Bellamy and increasingly evident in modern life is the seamless integration of desire, transaction and delivery, epitomized by the plans of retailers such as Amazon to develop drones for rapid delivery to your door. A similar model is described by Brian Aldiss as a feature of the fully automated city of the future, in his 1963 story "The Underprivileged", in which a credit card fitted into a slot at any street corner brings immediate delivery by robot drone.

ABOVE LEFT Cover of the May 1963 issue of *New Worlds Science Fiction*, where Aldiss's tale "The Under-Priveleged" [sic] first appeared.

ABOVE RIGHT Cover of *Starswarm*, the "fix-up" (a novel-length book created by stitching together previously published short stories) which incorporates the short story "The Underprivileged".

OPPOSITE Sample version of an early American Express credit card.

06

BIG BROTHER IS WATCHING YOU:
THE SURVEILLANCE STATE PREDICTED

George Orwell's novel *1984*, published in 1949, is celebrated as one of the greatest literary achievements of the sci-fi genre.

A bleak dystopian vision of a totalitarian future Britain, it was brewed from a complex mix of inspirations, including: the dismal shabbiness of postwar Britain; the stark truth, revealed by the excesses of the Second World War, of the capacity of the state and humanity itself for hideous repression and dispassionate evil; Orwell's disgust for the betrayed utopian ideals of Soviet socialism and its descent into Stalinist madness; his own experience of working on propaganda during the war; and his burgeoning pessimism about the evolution of human affairs. The book contains a remarkable range of foretold trends and technologies, from geopolitics to office technology, but it is best known today as a shorthand for oppressive state surveillance, or to put it another way, the rise of the surveillance state. But Orwell was not the only, nor by a long way the first, sci-fi writer to conjure visions of surveillance technology and the effects it might have on society. What is the relationship between fiction and fact in the present day?

PANOPTIC PASTS AND FUTURES

The signature surveillance technology of Orwell's *1984* is the telescreen, a wall-covering flat screen present in every home in Airstrip One, the new name for post-nuclear-apocalypse Britain. Dominating the living room of Orwell's protagonist Winston, the telescreen is described as "an oblong metal plaque like a dulled mirror which formed part of the surface of the right-hand wall". It functions both as television and surveillance

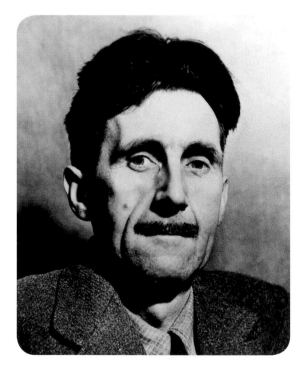

camera: "[it] received and transmitted simultaneously. Any sound that Winston made, above the level of a very low whisper, would be picked up by it, moreover, so long as he remained within the field of vision which the metal plaque commanded, he could be seen as well as heard." The telescreen constantly broadcast the ruling party's relentless propaganda, along with a steady diet of the most vulgar and brain-numbing entertainment. Perhaps most grimly, "The instrument could be dimmed, but there was no way of shutting it off completely."

ABOVE John Hurt as Winston Smith, seated in front of a telescreen in the 1984 film adaptation of Orwell's novel.

RIGHT Edmond O'Brien as Winston Smith in the 1956 film adaptation of *1984*.

OPPOSITE Portrait of George Orwell (real name Eric Blair) in 1946.

Watching through the telescreen are the Thought Police, the ever-present security force who are the instrument of repression. "There was of course no way of knowing whether you were being watched at any given moment," reflects Winston. "You had to live – did live, from habit that became instinct – in the assumption that every sound you made was overheard, and, except in darkness, every movement scrutinized."

When Orwell was writing *1984* televisions were still a novelty and hardly any homes had one. Within a decade they would launch their conquest of the domestic and cultural spheres, but Orwell was already looking beyond this, distorting the instrument of mass entertainment into one of mass repression. In this, as in many other aspects of *1984*, he was profoundly influenced by a little-known novel by Russian writer Yevgeny Zamyatin, titled *We* (written around 1921 but not translated into English until 1924). In the book, the world of the future is ruled by the One State, and the oppressed citizens live in a giant city almost entirely constructed from glass, so that the secret police can spy on everyone, all the time. Society is completely uniform and regimented, and the Bureau of Guardians keeps it that way.

The most obvious inspiration for Zamyatin's glass-walled urban prison-state is the Panopticon conceived by British philosopher Jeremy Bentham in the late eighteenth century. The Panopticon (from the Greek for "all-seeing") was explicitly designed as an instrument of social and psychological control. Although usually interpreted as a design for a prison, Bentham actually extended its purview to any form of institution, including hospital, asylum or school. In a Panopticon, cells, walkways and other rooms are constructed so that the walls and ceilings facing a central observation point are all transparent. A single "watchman" can thus, theoretically, observe any inmate at any time. Although a single person cannot watch everyone, the inmates will never know if they are currently under observation and so will be forced to police their own behaviour accordingly, exactly like the inhabitants of Orwell's Airstrip One.

Zamyatin's glass-walled city is a Panopticon on a colossal, urban scale. Also ubiquitous in the city are

ABOVE English translation of Zamyatin's influential novel *We*, which namechecks its literary heirs *1984* and *Brave New World*.

OPPOSITE Elevation, section and plan of Bentham's design for a Panopticon prison, in which every cell can be monitored from central watch stations.

"membranes": listening devices that "are handsomely decorated and are placed over all the avenues, registering all street conversations for the Bureau of Guardians". Although auditory in function, like Orwell's telescreens they are large, flat surveillance devices.

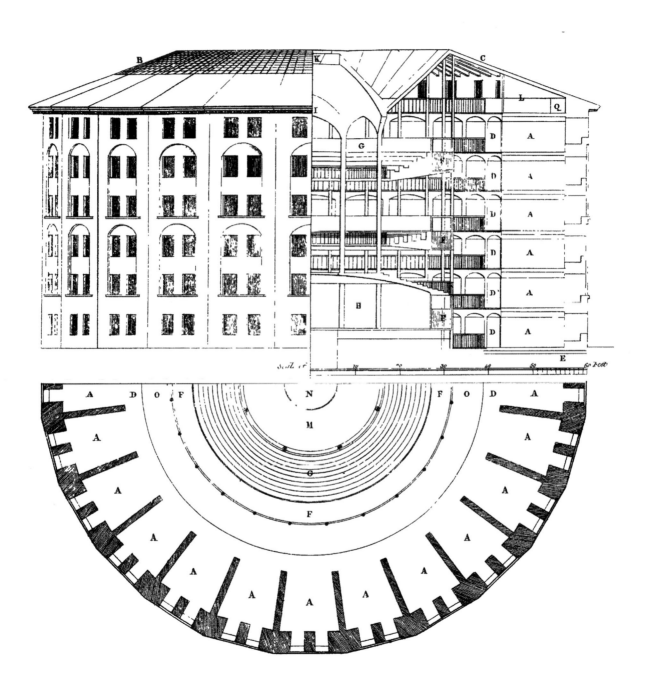

Telescreens and membranes are reminiscent of some of the latest and most concerning trends in surveillance: always-on, always-listening/watching internet-connected devices that people willingly install in their homes in the form of web cams, IP cameras and "smart speakers", such as Amazon's Echo or the Google Home.

ABOVE Cameras such as these are a common feature on British streets, silently monitoring the public.

OPPOSITE CCTV monitor grid showing multiple camera feeds, as predicted by Ray Cummings in 1939.

NATION ON CAMERA

The most commonly drawn link between Orwell's Big Brother state and real life is the spread of video surveillance, by means of what used to be known as close-circuit TV (CCTV) cameras; that is, cameras that only broadcast to a specific system not open to general view. Although the first CCTV cameras were invented, appropriately enough, by the Nazis to monitor V-2 rocket launches, they did not become commercially available until 1949, the very year of *1984*'s publication.

As instruments of state surveillance, CCTV cameras began to proliferate in the 1960s. In Britain, the first use of CCTV on public streets was in London's Trafalgar Square in 1960, as a temporary measure during a visit of the Thai Royal Family, with permanent installations beginning the following year at major London railway stations. Since then the United Kingdom has gone on to become the most video-surveilled nation on Earth. In 1969 there were just 67 police CCTV cameras in Britain; by the time Orwell's foretold year came around, rapid advances in the capacity of video storage had led to a boom in CCTV and the government was installing 500 new cameras every week. By 2013 the UK had an estimated one CCTV camera for every 11–14 people, and by

2018 over £2.2 billion a year was being spent on video surveillance systems, with nearly 6 million public and private cameras installed around the nation.

Although it seems highly likely that Orwell's Airstrip One would have made common use of such a technology, CCTV cameras are not explicitly featured and the telescreen seems something rather different (*see* above). But CCTV *was* foretold a decade before Orwell's novel, in astonishingly prescient detail, in a 1939 story titled "Wandl, the Invader" by American author Ray Cummings. In one passage the protagonists use what Cummings calls an "image-finder" to surveil a group in a café. In fact, the interior of the café is covered by a dozen image-finders, and the observers are able to switch between feeds on a "mirror-grid" (a bank of monitors similar to what would be found in any modern CCTV control room) to first find their targets and then zoom in:

In a moment our mirror-grid was glowing with the two-foot square image of the interior of the Red Spark Café ... Venza asked eagerly, "Which is he?" "Over there on the third terrace, to the left ... We'll get a closer viewpoint." The table in question was no more than a square inch on our image. "Here's your near-look." Our image shifted to another viewpoint.

An even earlier story – dating back to 1894 – predicted the use of video surveillance for traffic policing. In his 1894 book *A Journey in Other Worlds*, American tycoon, sometime author and most famous victim of the *Titanic* tragedy John Jacob Astor IV imagined life in the year 2000. He devoted much attention to the particulars of electric cars (see Chapter 8), and also detailed precisely how the speed cameras of the future (which he called "instantaneous kodaks") would work:

The policemen on duty also have instantaneous kodaks mounted on tripods, which show the position of any carriage [vehicle] at half-and quarter-second intervals, by which it is easy to ascertain the exact speed, should the officers be unable to judge it by the eye; so there is no danger of a vehicle's speed exceeding that allowed in the section in which it happens to be; neither can a slow one remain on the fast lines.

EYES IN THE SKY

The CCTV boom of the 1980s sparked a rash of Big Brother stories that now seem quaint in the light of subsequent developments in surveillance. Almost any such development you can think of was foretold by science-fiction authors, often a surprisingly long time ago. For instance, satellite photography has been used for reconnaissance and surveillance since 1959, and there are now dozens if not hundreds of such satellites in orbit (such missions tend to be classified). American author Jack Williamson described with precision the use of satellite photography for such purposes back in 1931, in his tale "The Prince of Space". His protagonist is shown a large print out of a photograph …

A photograph, taken from space, of part of the state of Chihuahua, Mexico. "And see!" He pointed to a little blue disk in the green-gray expanse of a plain, just below a narrow mountain ridge, with the fine green line that marked a river just beside it. "That blue circle is the first ship that came …"

Unmanned aerial vehicles (UAVs), aka drones, now perform the bulk of aerial reconnaissance and surveillance. Such devices, used for such purposes, have featured in multiple science-fiction works since at least the late 1920s. Ray Cummings's 1928 novel *Beyond the Stars* features perhaps the earliest description of a remote video "drone", although in this case it is a single-use projectile, which can be directed by use of an "invisible connecting ray" feeding into a "raytron apparatus".

Roger Zelazny's 1966 novel *This Moment of the Storm* features surveillance drones, called "Eyes", with lethal capability. The Eyes are semi-autonomous hovering surveillance drones equipped with cameras and guns. They work in swarms, controlled remotely from a central location. Zelazny's narrator describes how he "sent my eyes on their rounds and tended my gallery of one hundred-thirty changing pictures, on the big wall of the Trouble Center, there atop the Watch Tower of Town Hall … Then I put the auto-scan in full charge of operations and went downstairs for a cup of coffee". Zelazny could almost be describing modern UAV operators at work in their peculiarly mundane surroundings. Similar devices named "copseyes" were posited by Larry Niven in *Cloak of Anarchy* (1972). Described as golden and "basketball-sized", each has "a television eye and a sonic stunner [and] a hookup to police headquarters", and they are used to police designated "anarchy zones" where anything goes except violence: "The copseyes, floating overhead and out of reach were the next best thing to no law at all … Let anyone raise his hand against his neighbor, and one of the golden basketballs would stun them both. They would wake separately, with copseyes watching."

BUGS

Sci-fi stories also offer a glimpse into the likely near-future of such drone surveillance. As far back as 1936 American sci-fi author Raymond Z. Gallun described a bio-mimicking insect-like micro-drone in a story called "The Scarab": "… the Scarab buzzed into the great workroom as any intruding insect might, and sought the security of a shadowed corner. There it studied its surroundings, transmitting to its manipulator, far away now, all that it heard through its ear microphones and saw with its minute vision tubes." There have been multiple real-world programmes aimed at realizing precisely this kind of device. For instance, Robert Wood of the Wyss Institute at Harvard University has been working for over a decade on a project sponsored by DARPA (the Pentagon's advanced research funding programme) to

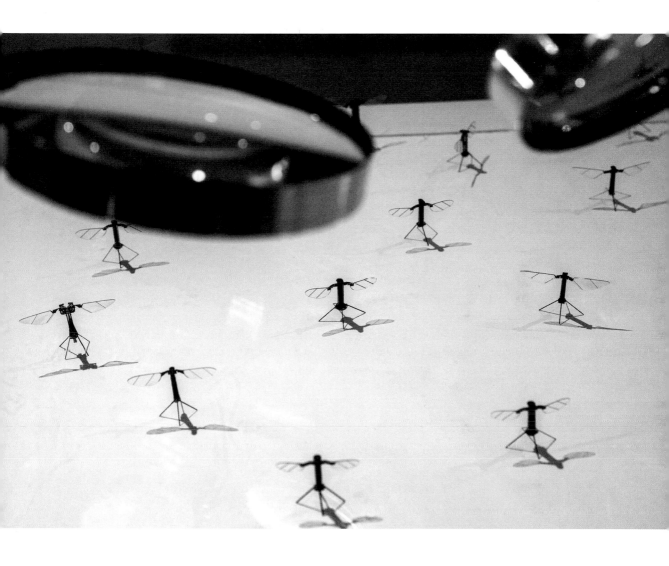

ABOVE Nanodrones known as RoboBees; technology like this could soon offer pervasive, ubiquitous and almost undetectable surveillance capabilities.

create an "autonomous flying microrobot", aka RoboBee.

A spin on the insectoid micro-drone that has been extensively pursued by DARPA-backed projects involves recruiting an actual insect to the cause and implanting electronics into it to create a cyber-bug. Such cyber-bugs might be remotely controlled via wires plugged into their nerves, while carrying cameras and listening devices. This approach lets nature and evolution take care of the difficult engineering challenges of building, maintaining and powering a tiny mobile surveillance platform. Examples include DARPA's Hybrid Insect Micro-Electro-Mechanical Systems programme; the cyborg beetle programme of Dr Hirotaka Sato of Nanyang Technological University in Singapore; and a North Carolina State University project to turn cockroaches and moths into disaster scene-searching "biobots".

But in the world of sci-fi, biobots are nothing new. Back in 1964 Philip K. Dick featured precisely this innovation in his novel *Lies, Inc.*, in which a character named Bill Behren, who describes himself as the "operator of fly 33408", boasts of its surveillance prowess: "... fly 33408 is a real winner. I mean it really gets in there and tackles its job and really gathers up the stuff, the real hot stuff. I've personally been operator for, say, fifty flies ... but in all this time, none has really performed true blue like this little fella."

TOTAL INFORMATION AWARENESS

The truly dystopian aspect of *1984* is not so much the surveillance technology used to spy on the citizens of Airstrip One, but the all-encompassing nature of Big Brother's intrusion. It is this chilling vision that makes the novel more relevant today than ever, as the space for personal privacy and civil rights shrinks ever smaller under the twin assaults of big business, with global corporations such as Facebook and Google mass-mining every aspect of personal data, and big government, with regimes from America to China using their security apparatus to achieve ever more ambitious data collection and control.

A key example is Echelon, a decades-long super-secret project co-ordinated by the US National Security Agency (NSA), Britain's Government Communications Headquarters (GCHQ) and other allied espionage services, and revealed in part by the leaks of Edward Snowden. Echelon was a broad-ranging programme to intercept electronic communications of all kinds, private and commercial, military and civilian. In the United States after 9/11, this mass eavesdropping evolved into something yet more sinister – a project called Total Information Awareness (TIA), which harboured the ambition of monitoring every communication everywhere, and employing algorithms and keyword filters to raise flags on any suspect interaction. The TIA programme was, ostensibly, shut down before it ever really got going, but suspicion about overreaching state

snooping apparatus remains widespread, reflected in films such as *Conspiracy Theory* (1997), *Enemy of the State* (1998) or the Bourne series of movies (including, for instance, *The Bourne Ultimatum* (2007), in which NSA assets are used to hunt for the eponymous super-spy). Such developments were foreseen by other sci-fi writers in addition to Orwell.

In John Brunner's 1975 novel *The Shockwave Rider*, for example, evil corporations exert global power through their control of data, and a rebel uses computer hacking to go on the run. In James Blish's *Cities in Flight* story series (started in the 1950s but first published in collected form in 1970) the City Fathers of the titular municipalities exert a terrifying level of surveillance, listening to every conversation – and responding in real time. In one passage two characters named Anderson and Chris are discussing the likelihood of something: "Anderson snapped a switch on his chair. 'Probability?' he said to the surrounding air. 'SEVENTY-TWO PER CENT,' the air said back,

OPPOSITE Bugging equipment used to spy on the Democratic headquarters is presented to the 1973 US Congressional enquiry into the Watergate scandal.

BELOW Archives of the East German internal intelligence agency known as the Stasi, which kept files on 10 million people.

making Chris start. He still had not gotten used to the idea that the City Fathers overheard everything one said, everywhere and all the time."

Using, presumably, the sort of AI algorithms that many suspect intelligence agencies are attempting to develop in real life, the City Fathers can even predict and remotely monitor police conversations. At one point Anderson begins to say something controversial: "'OVERRIDE,' the City Fathers said suddenly, without being asked anything at all. 'Woof! Sorry. Either I've already said one word too many, or I was going to. Can't say anything else, Chris … They are under orders

… to monitor talk about this situation and shut it up when it begins to get too loose …'"

The relentless self-censorship this implies harks back to the themes of *1984*, in which Big Brother seeks to achieve ultimate thought control through control of language itself: "It was intended that when Newspeak had been adopted once and for all and Oldspeak forgotten, a heretical thought … should be literally unthinkable."

Taking this grim system of oppression one step further, the citizens of Airstrip One are taught from infancy to internalize it and self-police their thoughts, a discipline known as CRIMESTOP in Newspeak:

CRIMESTOP means the faculty of stopping short, as though by instinct, at the threshold of any dangerous thought. It includes the power of not grasping analogies, of failing to perceive logical errors, of misunderstanding the simplest arguments if they are inimical to Ingsoc, and of being bored or repelled by any train of thought which is capable of leading

in a heretical direction. CRIMESTOP, in short, means protective stupidity ... orthodoxy in the full sense demands a control over one's own mental processes as complete as that of a contortionist over his body.

Compare this passage to reports of the "re-education" imposed on dissidents in the Soviet era, suspected bourgeoisie during China's Cultural Revolution or under Cambodia's Khmer Rouge, or Muslim Uyghurs in present-day China.

ABOVE Still from the 1998 film *Enemy of the State*, showing Will Smith being targeted by surveillance cameras as he attempts to avoid the clutches of the National Security Agency.

FUTURE SNOOPING AND DEEPFAKING

Alternative versions of TIA-style "total surveillance" have been posited by sci-fi writers exploring the concept of a time viewer, aka "chronoscope": a device that allows the user to see what is going on anywhere at any time, including the present or near-future. A version of this is the fundamental conceit of the 2002 film *Minority Report* (loosely based on a 1956 Philip K. Dick story "The Minority Report"), in which the existence of mutants able to view the future has led to the establishment of a law enforcement agency that arrests people based on crimes they are going to commit in the future. Other versions include Isaac Asimov's story "The Dead Past", also from 1956, in which the availability of "chronoscope" technology spells the end of privacy. Privacy is also killed off in Bob Shaw's 1972 novel *Other Days, Other Eyes*, in which even the tiniest fragments of a special glass can be used for surveillance and an oppressive state security system uses crop-dusting aircraft to saturate the environment with particles of the viewing medium. American sci-fi writer Damon Knight put a different spin on the same premise in his 1976 story "I See You", in which ubiquitous viewing technology has again eliminated privacy, but with positive, utopian results: humankind is liberated from all its hang-ups and neuroses and exists in a joyous state free from all vestiges of self-consciousness. Jeremy Bentham viewed his glass-walled Panopticon as "a mill for grinding rogues honest"; Knight expands the definition of "rogues" to include all of humanity, casting it as a liberation.

This optimistic interpretation of the looming death of privacy is squarely at odds with the bleak vision presented in Orwell's *1984*, summed up by its chief villain with his famous warning: "If you want a picture of the future, imagine a boot stamping on a human face – forever." Big Brother does not only want to change the future through omnipresent surveillance; he also has

RIGHT Jonathan Pryce as the Winston Smith equivalent character in the 1985 film *Brazil*, Terry Gilliam's dark comic adaptation of the novel *1984*.

designs on the past. In the novel, Winston's job is part of a society-wide programme to rewrite history and thus remodel reality itself, a task with clear parallels in the present day use of fake news and disinformation by malign actors or "bots" posing as humans, and right wing extremists. The relatively lo-tech media manipulation in which Winston engages – which involves some sort of speech-recognition, voice-directed editing technology – is likely to be exceeded in baleful potential by the looming marriage of disinformation campaigns with "deepfakes": AI-constructed fake videos that seamlessly merge different sources to produce footage of people saying or doing things they never said or did, but which is practically impossible to distinguish as fake.

ABOVE A Chinese police officer equipped with smart glasses, perhaps with facial-recognition technology.

LEFT Tom Cruise, Neil McDonough and Colin Farrell as future cops in the 2002 film *Minority Report*, where precognition allows surveillance of the future.

07

REPLICATORS IN ACTION:
STAR TREK AND 3D PRINTING

One of the most iconic recurring scenes in the much-loved TV series *Star Trek: The Next Generation* (*TNG*) involves the captain of the starship *Enterprise*, Jean-Luc Picard, alone in his cabin, announcing to thin air, "Tea. Earl Grey."

A machine set into the wall, resembling a large built-in microwave with its door missing, responds to this command by materializing a fine bone-china teacup on a saucer, filled with steaming hot tea. This remarkable device is a replicator (aka molecular synthesizer), a machine that can create almost anything out of thin air, but is particularly used for food and drink.

As with several other iconic *Star Trek* technologies, replicators are directly responsible for inspiring developments in real-life technology, which use 3-D printing to create food, meals, plastic and metal items, buildings and even quite complex machine parts (see below). *Star Trek* is far from being the only sci-fi source of inspiration for the dream of a device that can produce finished items from scratch.

TWO'S COMPANY

To trace the roots of *Star Trek*'s replicator it is necessary to understand that it is essentially a repurposed form of the transporter: the teleportation or matter transmission device that "beams" the crew between starship and planet surface. According to legend, the transporter was only invented because the original series lacked the budget to film special, effect-heavy scenes of planetary landing shuttles, but *Star Trek* did not invent the concept of matter transmission. Its first appearance in science fiction dates back at least as far as 1877, in Edward

Page Mitchell's story "The Man Without a Body", which prefigures George Langelaan's much better-known 1957 story "The Fly", by having a scientist suffer a teleportation mishap when his batteries run flat as he is only part-way through transmission, so that only his head rematerializes.

The replicator uses the same basic principle as the transporter, in which the atomic structure of a physical object is scanned and the information is used to reconstruct the object at the "receiving" end through energy-matter conversion. In practice, all transporters are replicators and matter "transmission" is a misnomer, because matter itself is not transmitted, only information. Every time Captain Kirk steps out of the transporter having "beamed up" from a planet's surface, it is in fact a copy of him – the original having been disintegrated during the initial phase of the operation.

This was precisely the mechanism of teleportation explored in one of the earliest stories on this theme. In Guillaume Apollinaire's 1910 story "Remote Projection", an inventor finds that his teleporter is actually a replicator and ends up with 841 identical copies of himself scattered around the world. This idea anticipated the well-known Teletransporter philosophy thought experiment by British philosopher Derek Parfit, which explored questions of continuity of identity. If a transporter is actually a replicator, is the Captain Kirk that steps off the transporter pad the same as the one that was "beamed up" from the planet? If the planet-side Kirk is not disintegrated in the process, and survives the process, which of the two Kirks is the "real" one? *Star Trek TNG* explored this precise scenario as an

ABOVE Scene from the original *Star Trek* series, with crew members at a bar equipped with a food replicator (on the wall at the left).

RIGHT A food replicator, in the influential "built-in microwave" style, from the original *Star Trek*, replicating glasses of wine.

ABOVE Crew of the Enterprise from *Star Trek TNG* materializing on the transporter pad, a device that raises many philosophical questions.

OPPOSITE Prototype of the Foodini 3-D food printer, widely compared to the *Star Trek* replicator.

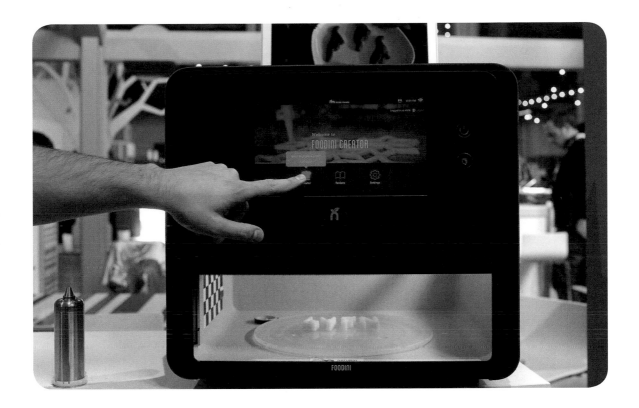

ongoing storyline, after an episode ("Second Chances", 1993) featuring a transporter malfunction that results in two copies of the character Will Riker – one who materializes on board his ship and the other who is stranded on a planet. The planet-side copy eventually chooses to be known as Tom Riker.

FOOD FOR THOUGHT

In the world of the TV series, the replicators of Picard's *Enterprise* are a development of food synthesizers: simpler machines present on James Kirk's *Enterprise* in the original *Star Trek* series (known as *The Original Series* or TOS). These closely resemble the later replicators, but were conceived by the writers of TOS more as highly advanced, mechanical food preparation devices, rather than matter-energy converters. They thus represented a televisual outing for a concept long popular in science fiction: the automatic food preparation device. A machine that performs complex autonomous physical

tasks can reasonably be described as a robot, and as early as 1899 Elizabeth Bellamy's novel *Ely's Automatic Housemaid* features a robot cook, which might be seen as a precursor to later food synthesizers.

Unspecified "mechanical apparatus" was at work in the automated canteens of Edgar Rice Burroughs's Mars. In his 1912 novel *A Princess of Mars*, Burroughs, probably inspired by the automat (a kind of vending-machine café imported from Germany to the US in 1902), describes "gorgeous eating places where we were served entirely by mechanical apparatus. No hand touched the food from the time it entered the building in its raw state until it emerged hot and delicious upon the tables before the guests, in response to the touching of tiny buttons to indicate their desires".

Moving from Burroughs's mechanical canteens to a *Star Trek*-style food synthesizer was simply a matter of miniaturization, and by 1933 David H. Keller was imagining "a small but complete production laboratory,

not much larger than [an] electric refrigerator … entirely automatic and practically foolproof". In his story "Unto Us a Child is Born", Keller envisages a machine that can both create food and prepare it "for the table in any form desired by the consumer. All that was necessary was the selection of one of the twenty-five menus and the pressing of the proper buttons".

Only recently has this dream of a kitchen appliance scale food synthesizer neared reality, with the launch of the Genie food replicator, which was explicitly inspired by *Star Trek*'s replicator. The Genie, a device not much larger than a microwave, with futuristic styling, claims to be "a kitchen in a box", which can make nutritious, freshly cooked meals in 30 seconds. However, it should be noted that the device relies on pods that contain dehydrated ingredients; in other words, the food preparation labour has simply been moved upstream in the process, and the Genie might be little more than a device for adding hot water to a dried-noodle pot.

IN DUPLICATE

What would it mean if a matter-energy conversion replicator were a reality? Matter duplication of the kind displayed by a replicator must inevitably cause extreme disruption of any possible economy. Such a scenario has been explored by sci-fi writers. George O. Smith wrote a series of tales, beginning with his 1945 story "Pandora's Millions", in which matter duplication makes all conceivable media of exchange worthless, until the discovery of identium, an element that cannot be duplicated, which then forms the basis for a new

ABOVE The Genie food replicator, marketed as real-life *Trek* tech.

OPPOSITE Cover of Damon Knight's *The People Maker*, in which replication technology immediately causes social breakdown.

ZENITH
BOOKS

35¢

LUST AND DECADENCE
RULED A WORLD
GONE MAD

DAMON KNIGHT

THE PEOPLE MAKER

A ZENITH ORIGINAL

ZB-14

currency. In Damon Knight's 1959 novel *The People Maker,* later published as *A for Anything* (1961), an inventor distributes several dozen copies of a Gismo that can duplicate anything, including itself. The only thing left of any value is human labour. Within weeks civilization breaks down and society mutates into a post-apocalyptic slave-based dystopia where warlords control the Gismos and duplicate their favourite subjects to create slave armies. A more optimistic take, in Ralph Williams's 1958 story "Business as Usual, During Alterations", envisages an America that happily switches from an economy based on scarcity to one based on abundance.

THE 3-D PRINTING REVOLUTION

The devices on show in *Star Trek* purport to work by converting energy into matter. This is problematic, because although Einstein's famous equation $E = mc^2$ shows that energy and matter are indeed inter-convertible, being simply two sides of the same coin, there is a considerable gap between this fundamental principle and any realistic prospect of human control of the process in the manner suggested by a replicator. For instance, the amount of energy needed to create a cup of tea is equivalent to about 6 megatons of TNT, or about 400 times the yield of the atomic bomb dropped on Hiroshima

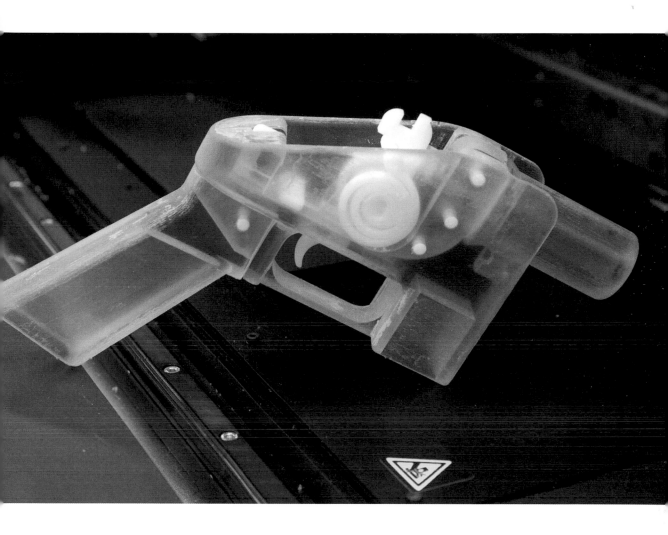

in 1945. Even if we allow that in the fictional future universe of *Star Trek* such energies can routinely be deployed, the picture is very different in the real world of today.

Currently, humans can just about manage conversion of infinitesimal amounts of matter into energy in the environs of a nuclear reactor or bomb. Running the process in the other direction can only be achieved in the collision chamber of a giant collider such as the Large Hadron Collider (LHC) at CERN in Geneva. In this rarefied environment, by accelerating particles to near-light speeds, a handful of sub-atomic particles can sometimes be smashed into fleeting existence: a long way from a cup of Earl Grey tea.

ABOVE A 3-D printed gun, modified to be safe, created with printed plastic parts.

OPPOSITE A MakerBot Replicator 3-D printer in action, deploying a plastic medium to build up a design, one layer at a time.

So what is meant by today's technology boosters when they talk excitedly about *Star Trek*-inspired, real-life replicators? The technology to which they refer is the 3-D printer: a device that lays down some form of plastic (in the sense of malleable) medium in layers to build up a three-dimensional form. Such printers are heralded as the drivers of a second Industrial Revolution, in which manufacturing is distributed and universal, available to all through desktop 3-D printing machines. These devices are already available, usually restricted to fabrication using quick-setting plastics or resins, but larger and more specialized machines can print in media varying from living cells and foodstuffs to metal to mud or concrete. Large-scale concrete printers, for instance, are suggested as a solution to housing crises such as those found in refugee camps where rapidly assembled, cheaply erected structures are needed. Meanwhile, biological implants and replacement tissues can be printed by laying down layers of cells on organic scaffolding, and in the near future it may be possible to print entire organs for transplant.

Although the 3-D printing community often plays up the lineage of inspiration from *Star Trek*'s replicators to desktop fabricators, in practice the former has a completely different mechanism. The true conceptual forefather of the 3-D printer is a 1964 story by Italian writer Primo Levi, "*L'ordine a buon mercata*" ("Order on the Cheap"). A mysterious multinational enterprise of dubious intentions makes available a device called the Mimer-duplicator, which can create exact replicas of anything from money and diamonds to food and humans. It works by extruding "extremely thin superimposed layers" of a multi-element substance named "pabulum". This is a concise and extremely accurate description of how a modern-day 3-D printer works.

ABOVE Italian author Primo Levi, whose 1964 story "Order on the Cheap" describes 3-D printing in precise detail.

OPPOSITE A concrete 3-D printer being used to print a house.

PART 3
SPACE & TRANSPORT

08

DRIVE TIME:
FROM ASIMOV'S "SALLY" AND
KNIGHT RIDER'S KITT TO THE DRIVERLESS CAR

The streets of Phoenix, Arizona, are currently playing host to a large-scale experiment that would have seemed absurdly futuristic just 10 years ago.

Cars are stopping to pick up passengers, pulling out into heavy traffic, negotiating busy streets, changing lanes, waiting for traffic lights to turn green, giving way to other vehicles, accelerating and braking in response to traffic flow and pulling over to drop off passengers. They are doing all these things with no input from a

driver, because these are driverless cars from the company Waymo (a subsidiary of Google's parent company). Waymo's "On the Road" programme boasts that, since 2009, its robot cars have driven well over 10 million miles: "We drive more in a day than the average American drives in a year".

Waymo is just one among a multitude of programmes and companies developing self-driving cars, although a host of different terms can be used: driverless cars, autonomous vehicles, pilotless automobiles, robot cars

and so on. At this point, whether or not they become an everyday commercial reality has as much to do with issues of regulation and public acceptance as technology. All three of these areas have been shaped by the legacy of driverless cars in science fiction.

KITT, JOHNNY AND CHRISTINE

For most people, the obvious sci-fi touchstones for self-driving cars are from film or television. Probably the best known is KITT, the co-star of the hit 1980s TV show *Knight Rider*, which teamed up with David Hasselhoff to fight crime and "champion the cause of the innocent, the powerless [and] the helpless". KITT was an artificial intelligence housed in the chassis of a customized 1982 Pontiac Trans-Am, with a distinctive personality: "Please do not refer to me as a 'car' or a 'set of wheels', it's most demeaning," it once told Michael Knight, the character

played by Hasselhoff, "I'm the Knight Industries Two Thousand. You [can call] me 'KITT'."

KITT had originally been hosted by a government mainframe but later its (his?) hardware (including a then-impressive but now slightly underwhelming one gigabyte of memory) was transferred to the Pontiac, which was tricked out with bulletproof armour and a distinctive "Anamorphic Equalizer" – a strip of LED

lights in its front bonnet that doubled as an optics array and mood indicator. It had highly advanced powers of smell and hearing, could mimic different accents, launch grappling hooks, oil spills and tear gas, and could even let its passenger play arcade games on the interactive screens in its interior.

An alternative vision of driverless cars was presented in the 1990 film *Total Recall*, directed by Paul Verhoeven and starring Arnold Schwarzenegger, based on the 1966 Philip K. Dick story "We Can Remember It for You Wholesale". In the film, the Schwarzenegger character Quayle is ferried around on Earth by "Johnny Cab", a taxi company comprising self-driving cars fitted with personable robot mannequins that make small talk and, theoretically at least, take instructions. Unfortunately, their natural-language processing is unreliable when dealing with vernacular, and they struggle to accommodate the demands of highly stressed-out passengers. Quayle becomes frustrated when the Johnny Cab fails to understand his

ABOVE View of KITT's interior, a 1980s-era vision of the future that failed to foresee the touch screen.

OPPOSITE Artwork from the poster for the 1983 comedy-horror film *Christine*, about a possessed 1958 Plymouth Fury.

command to just "Drive!" and he rips out the robotic mannequin in order to take control of the vehicle himself (which suggests that in fact the automobile is relatively conventional and that the robot itself was the driver). Johnny Cab thus offered a preview of the frustrations awaiting users of voice recognition services such as Siri and Alexa, and perhaps of issues that might arise with passenger-AI interfaces in the real-life autonomous vehicles of the future.

Distinctly alternative takes on the possibilities and pitfalls of automobiles with some degree of sentience

of the future in his 1894 novel *A Journey in Other Worlds*. He predicted not only the coming dominance of electric automobiles (which he called "phaetons") but also that they would be rechargeable, sketching out an entire system of renewable power generation with wind turbines and charging points. In assessing Astor's prescience, however, it should be noted that many of the early automobile designs were electric and that at the end of the nineteenth century it was by no means clear that the combustion engine would prove to be the dominant model for cars for at least the next century.

One of the earliest descriptions of a self-driving car, and the highly relatable motivations behind its invention, comes from a 1935 story by American psychiatrist and author David H. Keller, titled "The Living Machine". In the tale, an inventor named John Poorson is knocked into the gutter by a poorly piloted automobile. "And that," he concludes, "is just one more reason why the average human being should not be allowed to drive such a powerful machine!" Poorson develops "something new in the way of an automobile", and the scornful response of a fellow named Babson – "Nothing new about this" – is quickly silenced when he looks inside. "Where is the steering wheel?" asks a perplexed Babson. "I do not need one," proclaims Poorson, letting the machine guide itself into traffic. Keller even thought through some of the transformational possibilities of such technology for safety, personal mobility and making transport more accessible: "Old people began to cross the continent in their own cars ... The blind for the first time were safe. Parents found they could more safely send their children to school in the new car than in the old cars with a chauffeur."

were offered by films such as *The Love Bug* (1968) and *Christine* (1983). In the former, a Volkswagen Beetle named Herbie has a life and personality of its own, while in the latter, based on the Stephen King novel of the same name, a 1958 Plymouth Fury is possessed by a jealous and vengeful spirit that engineers various mishaps and murders. The animating sentiences in these cases are supernatural not sci-fi, but in the context of futuristic visions of cars that drive themselves, how much difference is there really between a discarnate sentience and a sentient computer programme? Not for nothing is the latter frequently characterized as "the ghost in the machine".

AUTONOMOUS AUTOS FROM ASTOR TO ASIMOV

KITT may be the most culturally accessible prototype of the driverless car, but science fiction was predicting advanced autos long before the Hoff got behind the wheel. Gilded Age tycoon and sometime sci-fi dabbler John Jacob Astor IV offered a vision of the automobiles

Notable examples of driverless cars occur in the fiction of sci-fi colossus Isaac Asimov. In his 1953 story "Sally", Asimov imagined a kind of retirement home for old autonomous vehicles, which appear to be equipped with personalities, akin to Herbie the Love Bug. In the world of this story, "automatics" were invented sometime around 2015, and were initially reserved for "blind war veterans, paraplegics and heads of state". The elderly narrator recalls with a shiver "when there wasn't an

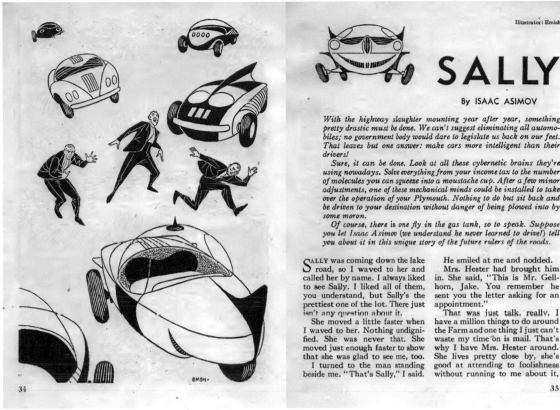

Illustrator: Emsh

SALLY

By ISAAC ASIMOV

With the highway slaughter mounting year after year, something pretty drastic must be done. We can't suggest eliminating all automobiles; no government body would dare to legislate us back on our feet. That leaves but one answer: make cars more intelligent than their drivers!

Sure, it can be done. Look at all these cybernetic brains they're using nowadays. Solve everything from your income tax to the number of molecules you can squeeze into a moustache cup. After a few minor adjustments, one of these mechanical minds could be installed to take over the operation of your Plymouth. Nothing to do but sit back and be driven to your destination without danger of being plowed into by some moron.

Of course, there is one fly in the gas tank, so to speak. Suppose you let Isaac Asimov (we understand he never learned to drive!) tell you about it in this unique story of the future rulers of the roads.

SALLY was coming down the lake road, so I waved to her and called her by name. I always liked to see Sally. I liked all of them, you understand, but Sally's the prettiest one of the lot. There just isn't any question about it.

She moved a little faster when I waved to her. Nothing undignified. She was never that. She moved just enough faster to show that she was glad to see me, too.

I turned to the man standing beside me. "That's Sally," I said.

He smiled at me and nodded.

Mrs. Hester had brought him in. She said, "This is Mr. Gellhorn, Jake. You remember he sent you the letter asking for an appointment."

That was just talk, really, I have a million things to do around the Farm and one thing I just can't waste my time on is mail. That's why I have Mrs. Hester around. She lives pretty close by, she's good at attending to foolishness without running to me about it,

ABOVE The preface and opening paragraphs of Asimov's "Sally", trailed as a story about what happens when "cars are more intelligent than their drivers".

OPPOSITE Isaac Asimov, author of a 1953 story about self-driving cars and a pioneer in the field of robot ethics.

automobile in the world with brains enough to find its own way home … Every year machines like that used to kill tens of thousands of people".

Asimov went on to predict what will surely be one of the most pressing constraints on the widespread adoption of driverless cars in the real world: the technology only truly works to its potential if all the

cars are autonomous, but the public may not stand for this. "I can remember … when the first laws came out forcing the old machines off the highways and limiting travel to the automatics. Lord, what a fuzz. They called it everything from communism to fascism." He also predicted some of the changes that such technology would bring to mobility habits, such as the rise of ride-sharing and the rarity of private individual car ownership. One of the more intriguing suggestions by real-world self-driving car futurologists, for instance, is that such vehicles might function as autonomous commercial enterprises, hiring themselves out and managing their own finances, so that no-one owns their own car any more because the cars own themselves.

Asimov saw robot cars as more than mere plot device. In a celebrated 1964 essay looking forward to what

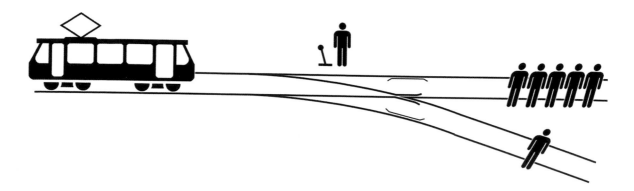

might be on show at the World's Fair of 2014 – in which he predicted ready meals, cordless appliances, large solar power stations in desert regions and the beginnings of consumer robot technology ("robots will neither be common nor very good in 2014, but they will be in existence") – Asimov outlined a vision of driverless cars: "Much effort will be put into the designing of vehicles with 'Robot-brains': vehicles that can be set for particular destinations and that will then proceed there without interference by the slow reflexes of a human driver."

TROLLEYS AND TUNNELS

Perhaps Asimov's most significant contribution to the development of self-driving cars ostensibly has nothing to do with vehicles. Asimov was the leading sc-fi thinker on the subject of robot ethics: the questions that genuinely autonomous and intelligent robots will raise in terms of how they relate to humans and each other. He drew up a kind of Three Commandments for robots with his celebrated Laws of Robotics:

> 1. A robot may not injure a human being or, through inaction, allow a human being to come to harm.
> 2. A robot must obey orders given it by human beings except where such orders would conflict with the First Law.
> 3. A robot must protect its own existence as long as such protection does not conflict with the First or Second Law.

ABOVE Graphic presentation of the Trolley Problem: should the person at the points switch the runaway trolley to the track with only one person?

Asimov was able to use fiction to work through the application – and loopholes – of these laws, but the designers, engineers and programmers behind driverless cars must face up to real-world versions of such challenges. The basic paradigm for exploring these issues has been a version of what philosophers call the Trolley Problem, after a thought experiment first proposed by British philosopher Philippa Foot in 1967.

The Trolley Problem concerns a runaway train car or trolley on rails, which is about to run over several workmen on the track, and asks whether it is ethically acceptable, or even imperative, to switch the points in order to send the trolley down a different track on which there is only a single worker. The default argument for working out the answer is the philosophical approach known as utilitarianism, which equates the ethically correct course of action with the one that maximizes benefit (known as utility). In the basic trolley problem this seems to suggest that it is ethical to switch the points as it is better for one person to be run over than several. Utilitarianism is one of the obvious approaches to use when trying to build an ethical dimension into machine intelligence, since it involves (theoretically) quantifiable variables amenable to calculation.

A version of the Trolley Problem involving a tunnel is one of the key paradigms for exploring the ethics of robot cars. It revolves around a scenario in which a self-driving car is heading at speed towards the mouth of a single-lane tunnel when a pedestrian darts into the middle of the road. The car's computer "brain" must make an instantaneous decision: should it swerve to avoid the pedestrian, in which case it will definitely hit the wall on either side of the tunnel, or should it drive straight on and run over the pedestrian? Should the moral calculus here be affected by considerations such as the number and identity of the pedestrians or passengers? Should the car have some sort of ethics setting that affects or constrains its decision-making, and if so, who should set it?

Rules such as Asimov's Laws of Robotics can help to guide the thinking of system designers in the real world. They provide a starting point, so that, for instance, driverless cars are likely to be programmed to prioritize human safety over that of the vehicle itself, but not to obey human orders if these will endanger human safety. Where the rules do not provide clear guidance, lead to paradoxes or cannot cope with vexing challenges, Asimov's fictional explorations of those grey areas may help to provide a roadmap for the real world.

ABOVE Illustration from 1957 imagining a family playing a board game while their car drives itself.

BELOW A driverless "solar bus", able to recharge as well as drive itself. Such a vehicle could become truly autonomous.

ROAD HOGS

For a vision of what it might look like if driverless vehicles have their ethical parameters set to "evil", or at least "indifferent", the 2017 film *Logan* offers clues. Set in a bleak dystopian America in 2029, the movie shows highways dominated by autonomous trucks carrying cargo containers at high speeds with little regard for other road users, including pedestrians. This is a version of the future in which the order of priority of Asimov's laws seems to have been reversed.

The autonomous trucks of *Logan* also bring home an important point about the driverless vehicles of the future: their design and aesthetics may be very different from those of today. Freed from the design imperatives

imposed by the need for drivers – such as visibility, dominant frontal positioning, and so on – autonomous vehicles can explore new set-ups, from the brutal functional aesthetic of *Logan*'s self-driving trailers, which are nothing more than stripped-back platforms for cargo containers, to the executive lounges on wheels found in many of today's concept driverless cars.

ABOVE Concept self-driving car. With no need for a driver, steering wheel or windshield, the interior of the car can be radically reconfigured.

09

SUBMARINES:
NAUTILUS AND ITS REAL-LIFE COUNTERPART THE *ARGONAUT*

Jules Verne formed one half of the twin foundations of science fiction. Along with H. G. Wells, he laid the template for what science fiction could be: exciting romances filled with gripping adventure, hinging on insightful grasp of cutting-edge science and technology and visionary grasp of the possibilities and ramifications of future technology.

Perhaps his most famous creation is the archetypal steampunk machine of scientific romance: the submarine *Nautilus*. Verne's vision of *Nautilus* and the underwater world of adventure to which it opened the door was informed by exciting contemporary advances in submarine technology, and would go on to be profoundly inspirational to the next generation of submarine designers. His novel would help to bring into existence the submarines that would wage – and nearly tip the balance of – the next war, and it is no coincidence that a string of submarines which share a name with his fictional version have seen active service. Many aspects of the *Nautilus*'s design and technology were highly original and foresightful.

AN ASTONISHING PHENOMENON
In his 1870 novel *Twenty Thousand Leagues Under the Sea*, Verne described a vessel that he clearly meant not to be fantastic but rather visionary. Through meticulous and detailed explanation, Verne implied that, although the vessel he described far outstripped contemporary marine technology in every way, it was nonetheless not just plausible but realizable, using technology then available. If only, Verne seemed to be saying, there were a daring person with the vision, intelligence, will

ABOVE Jules Verne, science-fiction visionary noted for his attention to detail.

OPPOSITE Captain Nemo uses a sextant while standing on the upper deck of the surfaced *Nautilus*. Sometimes old technology works best.

LEFT Aquanauts on an excursion from the *Nautilus*, using a system prefiguring SCUBA technology, marvel at the wonders of the ocean depths.

OPPOSITE The iconic Disney iteration of the *Nautilus*, from the 1954 film adaptation of *20,000 Leagues Under the Sea*, which has been a major influence on the steampunk aesthetic.

and resources, such a device could be constructed using existing science. In the vessel's proprietor, Captain Nemo, Verne presents such a person, and thus introduced into science fiction the archetype of the scientific superman; one who, through force of reason and willpower, can achieve technological marvels and realize impossible ambitions. As both superhero and supervillain, this archetype would be recreated throughout subsequent sci-fi and fantasy, producing such iconic figures as Superman's arch-nemesis Lex Luthor and Marvel Comics's "Iron Man", Tony Stark.

The novel provides relatively little description of the exterior appearance of the *Nautilus*. In his initial encounter with the vessel, the novel's narrator, Professor Aronnax,

believes it to be a giant marine animal until it comes to the surface: "There, a mile and a half from the frigate, a long blackish body emerged a metre above the waves ... This animal, this monster, this natural phenomenon that had puzzled the whole scientific world ... was, there could be no escaping it, an even more astonishing phenomenon – a phenomenon made by the hand of man: some kind of underwater boat that, as far as I could judge, boasted the shape of an immense steel fish."

In subsequent chapters, Captain Nemo gives Aronnax a tour of the interior (complete with luxurious library, museum and specimen collection) and a detailed account of its construction and workings. "It's a very long cylinder with conical ends", Nemo explains. "It noticeably takes the shape of a cigar, a shape already adopted in London for several projects of the same kind. The length of this cylinder from end to end is exactly

seventy metres, and its maximum breadth of beam is eight metres." This basic layout seems fairly conventional to modern readers, accustomed to the stereotypical cigar-shaped submarine design. However, in 1870 it was not at all clear that this was what a submarine "should" look like. Some contemporary experiments in submarine technology looked very much like sailboats; others were fat, barrel-shaped things. Verne's *Nautilus* helped to create the standard blueprint for the submarine.

ELECTRIC MARVELS

Verne also anticipated technologies that would become indispensable in real-world submarines. For instance, Nemo describes how he achieves buoyancy control using diving planes (fins that can be angled to act almost like aeroplane wings in reverse) and ballast tanks with compressed air: "I have supplementary ballast tanks

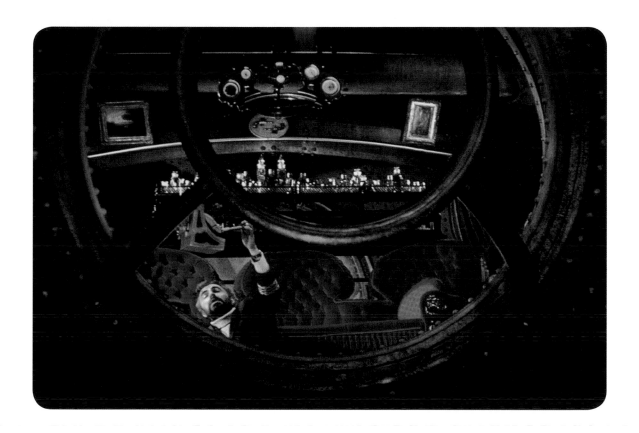

capable of shipping 100 metric tons of water. So I can descend to considerable depths. When I want to rise again ... all I have to do is expel that water."

To power his submarine marvel, Nemo uses electricity supplied by powerful batteries. At the time Verne was writing, contemporary submarine technology was struggling with the problem of running sufficiently powerful engines when fuel, air supply and exhaust must necessarily be restricted. The dominant marine engine technology of the age was steam, but this was clearly unsuitable for submarines. Electricity was still relatively novel and was yet to find widespread application outside the laboratory. However, Verne had the clarity of vision to see that electricity could be used to power all the different needs of a large vessel; that it offered a renewable energy source that could be recharged using elements abundant in the oceans, obviating the need for carrying bulky fuel or constantly putting into port to refuel;

ABOVE James Mason as Captain Nemo, in the 1954 film, amidst the splendour of his stateroom; in the novel it contains a collection of artworks and specimens.

OPPOSITE Illustration featured in the French edition of the book shows the engines of the *Nautilus*.

FOLLOWING PAGES Dramatic poster for the 1954 Disney film *20,000 Leagues Under the Sea*.

and that it could do all this with a minimum of noxious exhaust products. "It's marvellous," reflects Aronnax, "and I truly see, captain, how right you are to use this force; it's sure to take the place of wind, water and steam." The *Nautilus* uses electricity to drive its motors, and also to power electric lights, hobs and ovens, communication networks and all manner of other equipment.

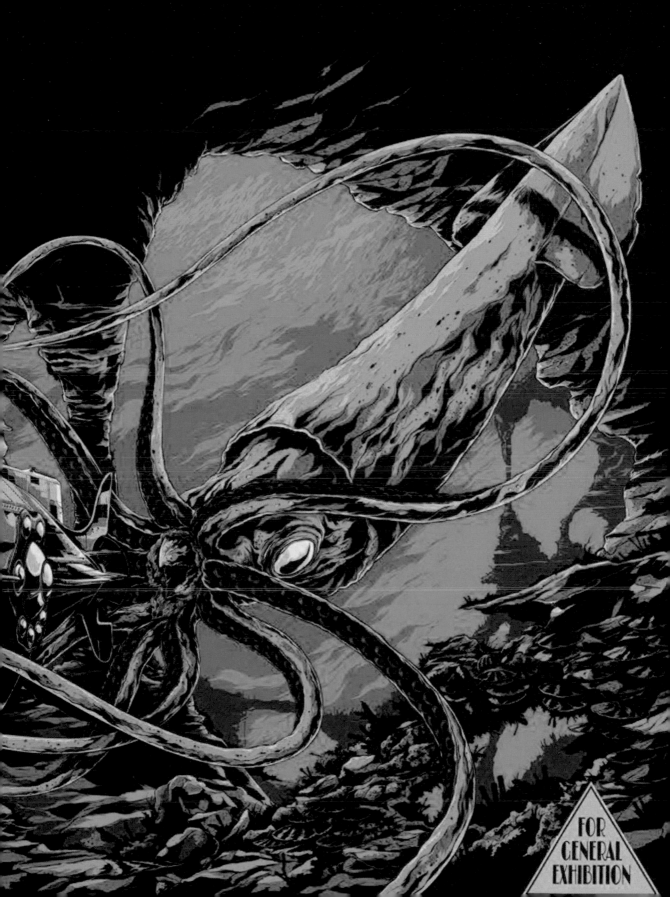

Another submarine technology that Verne seems to have anticipated was the scuba apparatus. Also known as the Aqua-Lung, scuba is actually an acronym for "self-contained underwater breathing apparatus". It was contrived in the 1940s by Jacques Cousteau and others, and frees divers from needing to maintain a connection to the surface by equipping them with their own compressed air supply and a tool known as a regulator, which safely decompresses the air for breathing. In *Twenty Thousand Leagues* …, Nemo describes a remarkably similar-sounding apparatus, complete with "regulator": a "tank built from heavy sheet iron in which I store air under a pressure of fifty atmospheres. This tank is fastened to the back by means of straps, like a soldier's knapsack. Its top part forms a box where the air is regulated by a bellows mechanism and can be released only at its proper tension." Nemo and his men were also equipped with compressed air rifles for underwater hunting, which shot miniature Leyden jars: electric shock bullets akin to tasers.

SUBMARINE DREAMS

The *Nautilus* may have been visionary, but Verne did not invent the submarine. At the beginning of the sixteenth century Leonardo da Vinci made cryptic notes and diagrams of some type of "ship-sinking-device" that has been interpreted as a submersible or submarine (the former is an underwater craft dependent on surface support, while the latter is fully autonomous; in general, however, it is acceptable to use submarine as an umbrella term). Like many of Da Vinci's inventions, his so-called submarine arguably belongs to the realm of speculative fiction, and it seems likely that he did not

OPPOSITE Sketches by Leonardo da Vinci showing aspects of submersible technology, widely over-interpreted as marking the invention of the submarine.

BELOW Engraving of the launch of one of the Winans' "cigar ships" — not submarines, but probably the inspiration for the shape of Verne's *Nautilus*.

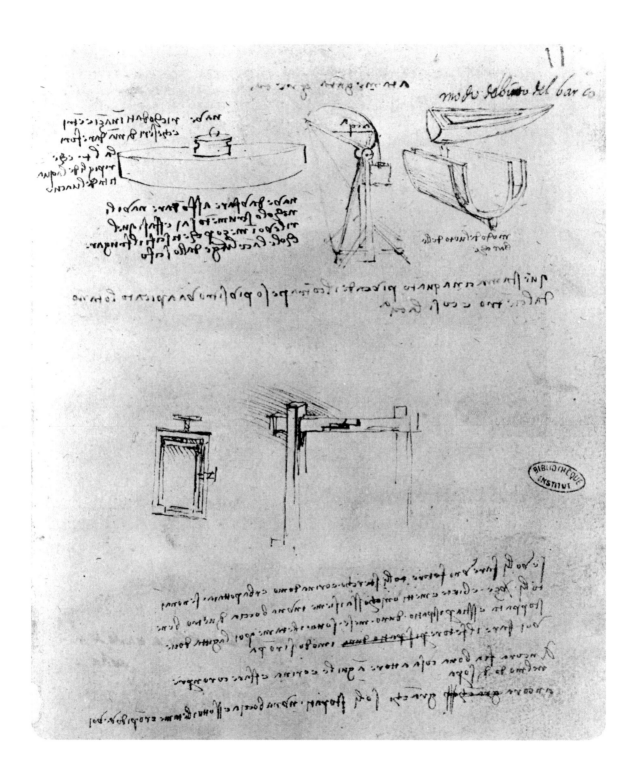

have in mind a true, fully submersible vessel. In 1578, English mathematician and sometime innkeeper William Bourne designed a submersible device that manipulated the principle of buoyancy to rise or sink in the water.

Around 1620, the Dutch inventor Cornelis Drebbel demonstrated in London a device that is often celebrated as the first submarine. It is even purported that King James I took a ride under the Thames in the vessel, and that in the course of inventing a means of air supply Drebbel also "discovered" oxygen. In fact, Drebbel's contraption was a kind of inverted boat: open at the bottom but closed overhead. It was powered by oars that projected through sealed ports along the side, and it seems unlikely that it could actually fully submerge, although it is hypothesized that it had a downward-sloping prow, so that rowers could propel it beneath the surface temporarily. It is almost certainly a myth that the King ventured on board the vessel, but it may well be true that Drebbel heated saltpetre (potassium nitrate) to produce oxygen in order to replenish his air supply. This does not, however, mean that he discovered oxygen – only that he made use of a well-known method for producing breathable "airs".

Perhaps the first person to write fully fledged submarine speculation was the English clergyman and natural philosopher John Wilkins. In his 1648 treatise *Mathematical Magick*, Wilkins discussed submarines, asserting their feasibility to be "beyond all question, because it hath been already experimented here in England by Cornelius Dreble". He went on to imagine submarine colonies, in which children would grow up without ever seeing land or sky, and discussed how food and waste might be passed into and out of a submarine.

Experiments with submarines continued without much success up until the end of the eighteenth century, when the first practical versions began to emerge. In 1797 Robert Fulton, an American artist and inventor living in Paris, offered the French authorities the services in battle against the British of what he called "a Mechanical Nautilus. A Machine which flatters me with much hope of being Able to Annihilate their Navy." Fulton eventually built his *Nautilus* and made numerous descents in it but

never achieved military success. The American Civil War prompted some famous submarine experiments as the Confederate side struggled to cope with a naval blockade, resulting in a vessel called the CSS *Hunley* (named after the cotton broker who supervised its design evolution and died during a test operation). In 1864 the *Hunley*, equipped with a "spar-torpedo" (an explosive charge on the end of a lance projecting from the prow), did contrive to sink a Union ship – the sloop *Housatonic* – but was lost with all hands immediately afterwards.

It was around the same time that father-and-son American marine engineers Ross and Thomas Winans developed a revolutionary and distinctive new ship design that it is believed to have provided the critical inspiration for Verne's concept of the *Nautilus* – even though it was for a surface vessel and not a submarine. In 1858 the Winans launched the first of their "cigar ships" – a vessel with a spindle-shaped design, the tapering prow and stern of which gave it a strong resemblance to a rolled cigar. Initially the ship was intended to be powered by a steam-driven paddlewheel set in the centre of the boat, which revolved around its exterior. This proved wildly impractical and was later replaced by propellers at each end, but the crucial element of the design was the extremely hydrodynamic hull, which was supposed to reduce drag dramatically and allow the ship to achieve high speeds with great fuel efficiency, while its sealed exterior made it invulnerable to most weather or sea conditions. In practice, the vessel could not adequately control its rolling and pitching and never achieved its potential, but it is likely that Verne saw one berthed in London (hence his reference in the text to the "cigar shape … already adopted in London for several projects of the same kind").

A CONTEST BETWEEN SUBMARINE BOATS

In the late nineteenth century, competition to attract naval patronage drove fierce competition between rival submarine designers, especially in the United States. At the forefront of this drive was American marine engineer Simon Lake, now eulogized by his supporters as "the father of submarines". At the age of 12, Lake had read

Verne's *Twenty Thousand Leagues Under the Seas* and been inspired to build in real life the marvellous vessel of the novel. His Lake Submarine Company of New Jersey built the *Argonaut*, which in 1898 became the first submarine to operate successfully in the open sea, making a passage from Norfolk, Virginia, to Sandy Hook, New Jersey, and prompting Verne himself to send a congratulatory telegram from France:

> While my book *Twenty Thousand Leagues Under the Sea* is entirely a work of imagination, my conviction is that all I said in it will come to pass. A thousand mile voyage in the Baltimore submarine boat (*The Argonaut*) is evidence of this … The next great war may be largely a contest between submarine boats.

ABOVE The *Argonaut* being built, c.1900. It bears the familiar cigar shape later adopted by all submarine designs.

RIGHT Marine engineer Simon Lake, "the father of submarines".

Once more, Verne proved his remarkable powers of foresight. On the eve of the First World War, few senior naval officers took seriously the concept of submarine warfare. The British in particular objected to the "dishonourable" nature of submarine warfare, and most navies considered that submarines would be restricted to operations in and around harbours. But the power of the submarine as a tool to counteract surface-vessel naval superiority and the resulting blockades that this made possible, first hinted at by operations in the American Civil War, nearly made it the crucial technology of the war. The Germans, pressed to the limit of survivability by the British naval blockade of the maritime trade on which their war effort and national diet depended, wholeheartedly adopted the potential of submarine warfare. By allowing their U-boat fleet to

conduct unrestricted warfare against merchant as well as military shipping, the Germans nearly succeeded in bringing Britain to her knees. German submarines sank ships faster than Britain could build them, and it was only the entry of the United States into the war, coupled with a belated adoption of convoy tactics, that allowed the German U-boat menace to be overcome.

OPPOSITE Nuclear submarine the USS *Nautilus* passes under New York's Brooklyn Bridge in 1958.

BELOW James Cameron's *Deepsea Challenger* submersible, the nearest latter-day equivalent to Nemo and his *Nautilus,* being lowered into the ocean for a visit to the uttermost depths.

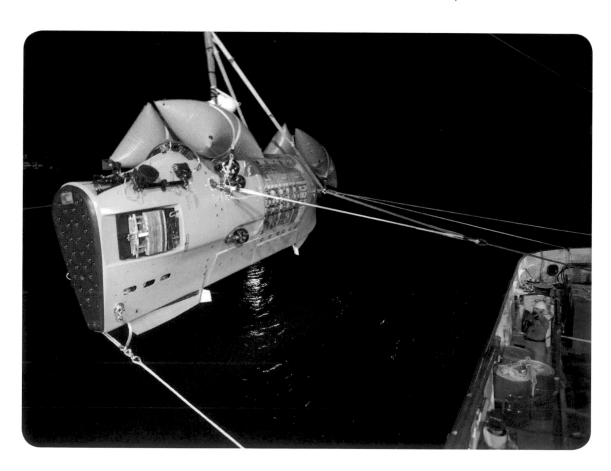

Verne's *Nautilus* continued to exert its influence. Submarines such as the US Navy's USS *Nautilus* – the first operational nuclear-powered submarine, which achieved the historic feat of sailing under the Arctic ice cap, thus becoming the first vessel to reach the geographic North Pole – continued to bear the name. Intriguingly, the USS *Nautilus* was launched in the same year that Disney adapted Verne's book and made the fictional *Nautilus* nuclear-powered as well, in the 1954 James Mason film *20,000 Leagues Under the Sea*. At the end of the movie Nemo's secret island lair detonates with a suggestively mushroom cloud-shaped explosion.

The legacy of Verne's *Twenty Thousand Leagues ...* is still influential today, from the pervasive steampunk aesthetic that refuses to fall out of fashion to the allure of deep ocean exploration. How much is owed to the influence of Nemo and the *Nautilus* by exploits such as those of James Cameron, the film-maker who developed a futuristic deep ocean submersible, the *Deepsea Challenger*, and piloted it solo to the deepest part of the Earth's oceans in 2012? Although Cameron's sub does not look much like Verne's *Nautilus*, the two have much in common: both were created by wealthy visionaries of saturnine character, both are powered by electricity and use revolutionary technology and both push the boundaries for deep-ocean exploration.

10

MOON ROCKETS IN FLIGHT:
JULES VERNE, TIN TIN AND THE MOON-SHOT

Outer space is the natural territory of science fiction, and the Moon was the destination of several of the first works of proto sci-fi.

German astronomer and mathematician Johannes Kepler's *Somnium* ("Dream"), written around 1608, is sometimes described as the first work of science fiction. It concerns a visit to the Moon, and although travel between the Earth and the Moon is accomplished with the aid of supernatural entities, Kepler included in the book cutting-edge science concerning the heliocentric solar system and orbital mechanics, hence the label "science fiction". The French soldier and writer Cyrano de Bergerac wrote a satirical work titled *Other Worlds: The Comical History of the States and Empires of the Moon and Sun*, which was published posthumously in 1657. By accident rather than design, its protagonist travels to the Moon by rocket, with the rocket boosters dropping away after accelerating him to sufficient velocity.

SHOOTING FOR THE MOON

Up until the mid-nineteenth century, Moon-shot tales failed to recognize the presence of a vacuum in outer space, although the traveller in Edgar Allan Poe's 1835 story "The Unparalleled Adventure of One Hans Pfaall" was perhaps the first to worry about his air supply. Poe was a major influence on Jules Verne, whose 1865 novel *From the Earth to the Moon* is regarded as the grandfather of serious moon-shot sci-fi. The book and its 1870 sequel *Around the Moon* concern a scheme to build a giant gun (the Columbiad) that will fire a projectile all the way to the Moon. Three men travel in the projectile, intending to land on the Moon, but a

close encounter with an asteroid causes them to deviate from their course so that they can only orbit the Moon. Eventually they use rocket propulsion to travel back to Earth and splash down in the Pacific Ocean. Although some elements of the tale are far-fetched and some slightly misguided, Verne worked hard to ground his tale in reality, calculating the forces involved and making a number of remarkably prescient predictions, albeit that some were probably coincidental.

Perhaps Verne's most impressive insight was to take serious note of the simple physical difficulties of escaping

the Earth's gravitational pull. Although he underestimated the force needed, he did at least attempt a calculation and correctly realized that colossal velocity would be required. Verne's moon-shot uses a massive cannon in the form of a shaft 270 metres (900 feet) deep and 18 metres (60 feet) wide, to accelerate a manned lunar capsule to escape velocity. In fact, the cannon he describes would not have been long enough, and even though he introduces a plot device to ameliorate the problem of the extreme acceleration (aka G-force) that the travellers would experience, in reality there would be no way to survive such forces.

Nonetheless, the list of accurate predictions made by Verne is impressive. He correctly identified the United States as the most likely nation to first achieve a moon-shot, and the novel's contest between Texas

ABOVE LEFT Frontispiece of the original French edition of Verne's 1865 novel *De La Terre à La Lune* (*From the Earth to the Moon*).

ABOVE RIGHT Illustration from Verne's novel showing the moon capsule's splashdown in the Pacific, remarkably close to where the real Apollo 11 capsule landed.

OPPOSITE Comical depiction of Cyrano de Bergerac en route to the Moon: perhaps the first rocket-propelled ascent in fiction.

and Florida as to which would host the launch mirrored the real-life competition between those states that eventually saw NASA site the launch pad in Florida and Mission Control in Houston, Texas. Verne's moon-shot is launched from close to Cape Canaveral, site of the actual Apollo launches, and the lunar projectile slightly resembles the actual Apollo command and service module craft. In both the real and fictional versions the moon-shot was made by a three-man crew, and both the real and fictional craft were equipped with "retro-firing" rockets intended to decelerate the craft so that it could land safely. Verne equipped his lunar capsule with life support, sealing it against the vacuum of space, heating it and employing chemicals for oxygen generation and waste-gas absorption. He correctly had his astronauts experience weightlessness (although he incorrectly believed it would only be experienced at the point where the gravitational pull of Earth and Moon precisely counteracted one another), and wrote with foresight about asteroids being captured by the Earth's gravity and exerting their own gravitational influence. His astronauts orbit the Moon and experience the abrupt change in temperatures associated with lunar day and night, and

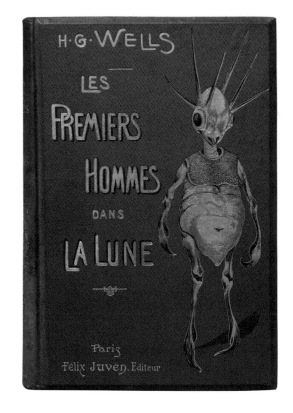

ABOVE French edition of H. G. Wells's novel *The First Men in the Moon*, graced by a Selenite, one of the insectoid lunar denizens.

OPPOSITE Poster for the 1964 film adaptation of Wells's novel, featuring Dynamation (stop-motion special effects created by Ray Harryhausen).

LEFT Still from *First Men in the Moon*, showing the lunar astronauts in deep-sea diver-style spacesuits.

H.G. WELLS' ASTOUNDING ADVENTURE IN
DYNAMATION!
H.G. WELLS'
COLUMBIA PICTURES
presents a
CHARLES H. SCHNEER
production
FIRST MEN IN THE MOON
IN PANAVISION AND LUNACOLOR!
starring
EDWARD JUDD · MARTHA HYER and LIONEL JEFFRIES
Screenplay by
NIGEL KNEALE and JAN READ
Associate Producer RAY HARRYHAUSEN · Directed by NATHAN JURAN · AN AMERAN FILM

they look down on a barren, lifeless landscape quite at odds with the traditional fantasies of lunar life but in line with reality. As with the real Apollo 11, upon their return to Earth Verne's astronauts splash down in the ocean, in almost exactly the same spot.

The carefulness and accuracy of Verne's lunar narrative stand in contrast to its most obvious point of comparison, H. G. Wells's 1901 novel *The First Men in the Moon*. Although Wells did feature the vacuum of space and weightlessness, he resorted to a fantasy device to get his travellers to the Moon (a fictional anti-gravity substance named cavorite), and he gave his Moon an atmosphere and an ecosystem, ruled by insectoid aliens. Verne chided Wells for his resort to the fantastical.

TO PENETRATE THE BOUNDS OF THE ATMOSPHERE

Around the same time that Verne and Wells were concocting their lunar narratives, the Russian scientist Konstantin Tsiolkovsky was writing the first genuinely realistic moon-shot sci-fi. Having written scientific papers about the use of rocket propulsion for spaceships, in particular his 1903 paper "The Probing of Space by Means of Jet Devices", Tsiolkovsky penned a science-fiction book for younger readers titled *Vne zemli* (1916; later translated as *Beyond the Planet Earth*). These works helped spread his message that the best route into space was via the use of multi-stage rockets. A rocket is a missile that travels by expelling material in one direction in order to generate a motive force in the opposite direction. Generally this is done by burning fuel to generate hot, expanding gases, which are expelled at the bottom of the rocket, thus driving it upwards. (A rocket that expels matter wholly or partly taken in from the surrounding space is called a jet.) Rockets have advantages such as being able to generate acceleration over sustained periods, so that they can build up speed, and they carry their fuel with them, becoming lighter and thus faster as they burn it up. As noted above, Cyrano de Bergerac had employed the device of a rocket for lunar travel, while John Munro's 1897 story *A Trip to Venus* also involved discussion of interplanetary rocket propulsion (see page 142). By 1911, American rocketry pioneer

Robert Goddard had begun experimenting with liquid-fuel rockets, inspired by his interest in science fiction. Tsiolkovsky's novel involved an international engineering programme to build rockets, and predicted the construction of orbiting space habitats and eventually the exploration and colonization of the solar system. His funerary monument carried the inscription "Man will not always stay on Earth; the pursuit of light and space will lead him to penetrate the bounds of the atmosphere, timidly at first but in the end to conquer the whole of solar space."

In the period between the First and Second World Wars there was a tight synergy between the science and science fiction of rocketry and spaceflight. In the Soviet Union a group led by Tsiolkovsky developed further his ideas about rockets, while a similar group in Germany saw spaceflight visionaries such as Hermann Oberth and Max Valier, and later Willy Ley and Wernher von Braun, team up to develop the ideas

К.Э.ЦИОЛКОВСКИЙ

Г Р Е З Ы

О·ЗЕМЛЕ И НЕБЕ

ОНТИ
1935

OPPOSITE Konstantin Tsiolkovsky, the Russian scientist lauded as the "father of rocketry and astronautics".

LEFT The cover of a Tsiolkovsky book about lunar exploration: note the inappropriate workwear of the lunar visitors.

behind rocketry and spaceflight, while also contributing to the science fiction literature that fed their dreams and inspired others to join them. Austrian engineer Valier, for example, wrote a technically detailed account of interplanetary rocketry in his 1928 story "A Daring Trip to Mars" (see page 142), and in 1930 became the

first man to die in the course of rocket research when he blew himself up with a liquid-fuel rocket. In 1928 Oberth was hired as a technical adviser by German film director Fritz Lang, and built a rocket for the pioneering sci-fi drama *Woman in the Moon* (1929). This landmark movie was the first true example of a sub-genre of sci-

fi films known as "spacesuit films", for their attempts to adhere to scientific reality, epitomized by the need to wear spacesuits to survive the vacuum of space (and in contrast to fantastical space operas such as *Flash Gordon*, which made little attempt to acknowledge the science in science fiction). Lang's film would have a direct impact on the future of space rocketry, introducing the launch countdown that would later be used in real space rocket launches.

DESTINATION MOON

The Second World War dramatically accelerated rocket research, as both the Allies and Germany raced to develop missiles. These were also golden years for science fiction, especially in America, where a young Robert Heinlein had teenagers build a rocket and fly it to the Moon to battle Nazis in his 1947 novel *Rocket Ship*

Galileo. Three years later he ditched the most fantastical elements of this tale and reworked it into the script of a hit movie, *Destination Moon*, which would present a hard science account of a moon-shot programme. The film featured face-contorting high G-factors at launch, spacesuits and spacewalks, and was a box-office hit.

Heinlein's film lent inspiration – and its title – to a book by the Belgian comic book author Hergé. His

ABOVE Models of Hergé's characters Tintin and Professor Calculus (Tournesol in the original French) in their spacesuits.

OPPOSITE Front cover of the Tintin adventure *Objectif Lune*, showing the rocket ship on its launch gantry. The form of the ship derives from the German V-2 rocket, while the distinctive colour scheme comes from US test rockets.

- HERGÉ -

LES AVENTURES DE
TINTIN

★

OBJECTIF LUNE

CASTERMAN

Tintin story *Objectif Lune* (translated as *Destination Moon*), was serialized in the same year (and published in book form in 1953) and featured a host of extraordinarily accurate depictions of elements of a moon-shot programme. The story concerns the young reporter Tintin, who agrees to be one of the astronauts aboard a nuclear-powered rocket designed by his friend Professor Calculus. Hergé went to great lengths to get the technical details right, taking advice from experts and doing lots of research, and his story and pictures display many accurate predictions and features. The rocket itself is closely modelled on the German V-2 rocket – the first intercontinental ballistic missile – which would form the basis for postwar American rocket development. It sports a distinctive red and white chequered livery, which Hergé copied from the colour scheme used by NASA in its testing programme to help measure the motion and rotation of rockets on launching. The rocket's launch gantry was copied from that used at the American White Sands rocket testing base, while the nuclear power plant was modelled on the American facility at Oak Ride in Tennessee – the nuclear pile within was depicted with remarkable accuracy. The ribbed spacesuits worn by the astronauts were based on existing models, as were the distinctive ergonomic beds that the astronauts lie on to help them cope with the tremendous G-forces experienced on lift-off. In the follow-up book, *Explorers on the Moon*, Tintin and friends use small rockets to turn the space vehicle around and then fire the main thruster to touch down softly. The Moon is depicted as an airless, dusty, rock-strewn and crater-pocked monochrome wasteland, very similar to the reality that the Apollo astronauts would encounter. The lunar explorers use a moon buggy and discover ice on the Moon – a contentious theory at the time Hergé was writing, but confirmed in the 1960s.

DRY AS MOON DUST

In summary, just how significant was the impact of sci-fi on the moon-shot programme? For many of the engineers and scientists behind the Apollo programme

that landed men on the moon, lunar spaceflight science fiction provided the cultural inspiration that drew them to the field and drove their pursuit of ambitious goals. More specifically, rocket scientists such as Tsiolkovsky and Max Valier used their fiction to help work through some of the basic science of rocketry and spaceflight, and hence these stories became integral to the development of the field. Later moon-shot fiction, such as that of Heinlein and Hergé, shared common ancestors with moon-shot science fact, so that both were drawing on the same sources. One intriguing line of descent from sci-fi to sci-fact concerns the recruitment strategies used by American aerospace and engineering firms in the 1950s, as they marketed themselves to the talent they would need in order to cash in on the emerging space race. As documented by archivist Megan Prelinger in her 2010 book *Another Science Fiction*, companies such as Northrup, Ex-Cell-O, Martin and National used science-fiction tropes to catch the attention of engineers and science graduates.

Prelinger, however, argues that the influence of science fiction may ultimately have hindered the progress of space exploration by raising unrealistic hopes and unrealizable visions that led to disappointment and disillusionment in the face of the reality. "The 'science fiction' aspect of future visualizations," she says, "grossly overstepped reality by too swiftly suggesting a land-based model of colonization where such a model just could not operate." Even today a complaint of the sci-fi community is that the space programme became bogged down in a "moon-dust" model (that is, more interested in boring science about moon-dust than inspiring exploration – what Prelinger calls "a schism between space for science, by scientists, and space for exploration, by explorers") and compromised by its grubbier aspects as part of the military-industrial complex. There is even a movement in current sci-fi,

OPPOSITE Destination Moon: Apollo 11 launches from Cape Canaveral on its historic mission to the Moon, following in the footsteps of Tintin.

exemplified by the Hieroglyph Project of American sci-fi author Neal Stephenson, to recreate the imagined golden age of cross-pollinating sci-fi and big science by producing avowedly inspirational sci-fi. In this context, the influence of sci-fi on the Apollo programme is understood to be the high watermark of science fiction directing the course of science fact. Certainly this was the view of spaceflight pioneer Wernher von Braun, who proclaimed of interplanetary science fiction, "It is the vision of tomorrow which breeds the power of action."

ABOVE Mission Operations Control Room at Houston, Texas, during the Apollo 10 mission. Launched on 18 May 1969 on a lunar orbital mission, Apollo 10 was the dress rehearsal for the Apollo 11 Moon landing.

OPPOSITE Buzz Aldrin on the surface of the Moon, photographed by Neil Armstrong, who can be seen reflected in Aldrin's visor.

11

INTERPLANETARY TRAVEL TO MARS:
WERNHER VON BRAUN, PROJECT MARS AND THE ELON CONNECTION

Almost as familiar a celestial body as the Moon, Mars has long attracted the imagination of writers and dreamers, and engendered in the world of science fiction at least two specific sub-genres of its own.

But one work of science fiction in particular – Wernher von Braun's 1949 tract *Project Mars: A Technical Tale* – has increasingly come to be seen as an amazingly prescient portent of a fictional future on the verge of becoming reality, from the precise technical details of its interplanetary travel to the unusual name of its Martian ruler: Elon.

LIFE ON MARS

In 1877 the Italian astronomer Giovanni Schiaparelli caused a sensation with his observations of Mars using the latest high-power telescopes. What particularly grabbed public attention was his use of the term *canali* to describe features on the planet's surface (features that would later turn out to be optical artefacts – illusions created by the process of imaging the Martian surface). This word meant simply channels, and hence was relatively neutral in implication, but it was widely translated into English as "canals", prompting a widespread belief that some intelligent civilization must be at work on Mars, engaged in planetary-scale civil engineering. The myth of the Martian canals was hyped by the work of other astronomers, most notably the Frenchman Camille Flammarion and the American Percival Lowell. Lowell built his own observatory in Flagstaff, Arizona and from 1894 made detailed observations of the Martian canals. Over the next two decades he developed an elaborate narrative (detailed

in books including *Mars* (1895), *Mars and its Canals* (1906), and *Mars As the Abode of Life* (1908)) explaining the canals as the work of an ancient Martian civilization attempting to irrigate its arid planet by transporting water from the poles in colossal quantities. This indigenous terraforming intriguingly foreshadowed what would become an important theme in later Martian science fiction.

With expert opinion disposed to consider Mars to be not only equipped with an atmosphere and capable of supporting life, but actually home to intelligent life and an

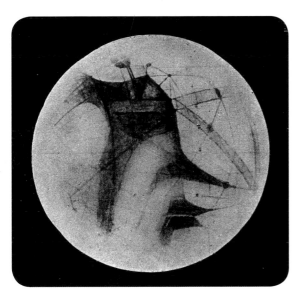

ABOVE One of Percival Lowell's drawings of the Martian surface, showing planetary-scale canal networks.

OPPOSITE Wernher von Braun, German-American rocket scientist and interplanetary spaceflight visionary.

RIGHT Camille Flammarion in his observatory at Juvisy-sur-Orge, near Paris.

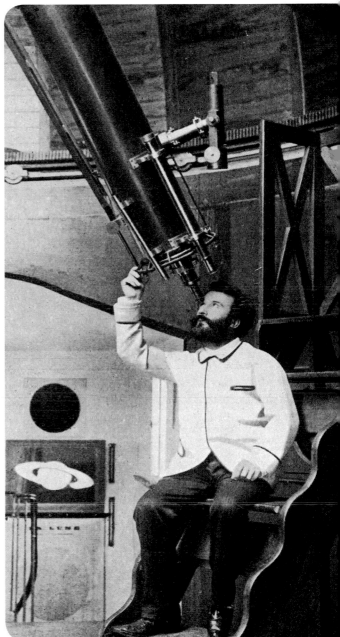

advanced and potentially ancient civilization, science-fiction writers had rich soil in which to plant their ideas. The best known is H. G. Wells's 1897 novel *The War of the Worlds*, in which technologically superior Martians seek to flee their dying planet and colonize the Earth. The picture of Martian civilization as exhausted and dying would linger on in sci-fi until well into the 1960s.

Drawing inspiration directly from Flammarion, the American adventure and sci-fi writer Edgar Rice Burroughs launched his massively influential "Barsoom" series of works set on Mars with his 1912 story "Under

ABOVE The glories of ancient Martian civilization,
depicted in *John Carter*, the 2012 Disney film
adaptation of Edgar Rice Burroughs' Barsoom series.

LEFT Edgar Rice Burroughs, prolific pulp-fiction author and larger-than-life character.

BELOW Classic cover art from Burroughs' first Barsoom novel, *A Princess of Mars* (a fix-up of stories originally serialized in 1912).

OPPOSITE Poster for the 1938 film *Flash Gordon's Trip to Mars*, showcasing all the elements of a classic space opera or "planetary romance".

the Moons of Mars", expanded to become the novel *A Princess of Mars* in 1917. In Burroughs's book, an American Civil War veteran, John Carter, is transported to the planet Mars (aka Barsoom) where he discovers a fantasy world peopled by strange, many-armed monsters, dastardly villains and beautiful princesses. Weaker Martian gravity gives Carter superhuman strength, and he pursues swashbuckling adventures across the landscape of the ancient planet and its decadent empires. Burroughs went on to write a further 10 Barsoom volumes over three decades.

SERIOUS JOURNEYS

Wells transported his Martians to Earth by means of colossal cannons not dissimilar to Verne's Columbiad (see page 124) but Burroughs fudged the crucial aspect of interplanetary transport by invoking some form of paranormal astral travel. However, a more technically grounded, "nuts-and-bolts" approach to the problem existed alongside the Burroughsian fantasy. For instance, Charles Dixon's 1895 "scientific romance" *1500 Miles an Hour* was deemed by the London *Spectator* to be "a serious narrative, as far as journeys to Mars and such things can be serious". It concerns a trip to Mars in a spaceship designed and built by one Dr Hermann, who describes it as "an air carriage, propelled by

electricity, capable of being steered in any direction, and of attaining the stupendous speed of fifteen hundred miles per hour". The book assumes the existence of a tenuous atmosphere in interplanetary space, so that the spaceship uses propellers. It features a very early example of a spacewalk (to fix a problem with the ship while en route).

John Munro's 1897 novel *A Trip to Venus* presents a dialogue in which a programme for interplanetary travel is discussed in some detail, including the use of rockets to propel and guide the spaceship to another planet. It features a detailed description of what would today be called a railgun – a device for electromagnetically accelerating projectiles. In the novel the idea is to generate thrust by reaction force through propelling projectiles from the rear of the spaceship; this exact principle forms the basis of modern ion drives, which use atomic-scale projectiles to generate thrust.

Scientific and technological advances in the early twentieth century – such as the rocket science of Goddard, Tsiolkovsky and Oberth (see page 128) – tended to constrain some of the more fantastical elements of Martian sci-fi. German rocket scientist Max Valier's 1931 story "A Daring Trip to Mars" is a technologically plausible account of a dangerous voyage to the Moon and Mars, in which lack of fuel prevents the astronauts from landing and they can only orbit the red planet. By the time Burroughs came to write his 1930s series *Carson Napier of Venus*, the science of interplanetary flight had become well known enough that Burroughs felt compelled to have his hero travel by a two-stage rocket with parachute landing, with a course and velocity determined by planetary fly-by/gravitational slingshot.

PROJECT MARS

Perhaps the most prophetic piece of Martian sci-fi was written in 1948 by the architect of the US space programme, Wernher von Braun. During the Second World War, Von Braun had overseen the development of the V-2 rocket programme for the Nazis; at the end of the war he, many colleagues and much of their research and materials were snaffled up by the Americans in the

infamous "Operation Paperclip". Relocated to the United States, von Braun became an enthusiastic American, a lynchpin of the military-industrial project and a key figure in the realization and marketing of the massive national effort to match the Soviet rocket programme and reach the Moon first. With Walt Disney's help, von Braun was repackaged into a national treasure, an avuncular figure who spread the gospel of science and engineering to kids across America in hugely popular TV programmes. Before all this, however, after the completion of the US Army's V-2 testing programme in 1948, von Braun found himself stuck in a US military base on the border of Texas and New Mexico with time on his hands and set himself to writing an account of his grand vision for interplanetary exploration. To make the mass of jargon and calculations more accessible, von Braun decided to use "an unpretentious tale as a frame in which to paint the picture. The idea has been therewith to beguile the tedium which might be caused by the relative dryness of disquisitions concerning each problem in detail".

The resulting book, *Project Mars: A Technical Tale*, though finished in 1949, was not published until many years later. But von Braun recapitulated its main points in a series of articles in the 1950s and published the book's technical annex in German as *Das Marsprojekt* in 1953 and in English in America in 1962 as *The Mars Project*. The original book relates in great detail an immense global programme, undertaken in the 1980s by a post-Second World War world government, to reach Mars. Von Braun's plan for the Martian expedition was to send 70 astronauts on a fleet of 10 large spacecraft, which would insert into orbit around the planet. The fleet would carry supplies and fuel to allow landing craft to touch down, explore, take off and rendezvous with the orbiting craft, seven of which would return to Earth. Given the mass of spacecraft, fuel and other supplies this would require,

OPPOSITE Cover art for the 1953 German publication of the technical appendices of von Braun's earlier *Project Mars*, published under the title *Das Marsprojekt* (*The Mars Project*).

Wernher von Braun

Das Marsprojekt

Studie einer interplanetarischen Expedition

Ein Sonderheft der Zeitschrift „WELTRAUMFAHRT"

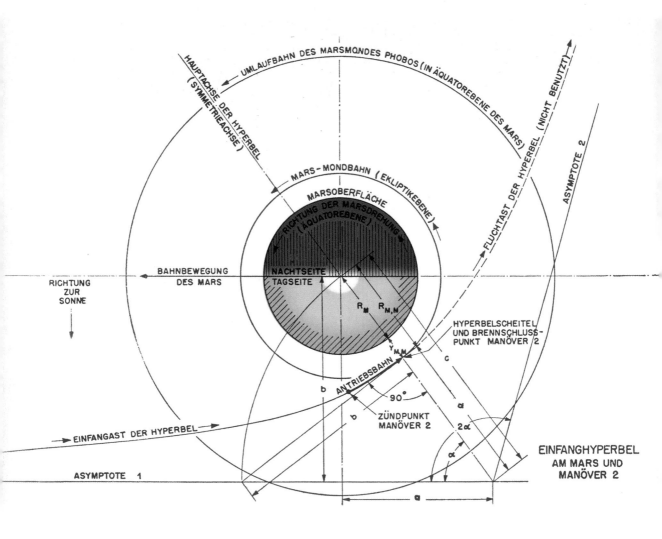

von Braun set out a programme for 950 Earth-to-orbit launches, using three-stage launch vehicles. Each of the three stages would be recoverable and reusable: the first two descending by parachute and the upper stage being a space shuttle that could glide to land on a runway. The three parts would be quickly refurbished and reassembled, allowing for a rapid and affordable cadence of launches. In this way thousands of tons of material could be lifted into orbit, where the interplanetary spacecraft fleet would be assembled and loaded.

In many respects this programme corresponds to modern plans for a crewed Martian expedition, particularly those set out by tech entrepreneur and industrialist Elon Musk. Musk's SpaceX is the highest profile and most successful private space launch company, revolutionizing the industry and radically reducing the cost of spaceflight by making increasing proportions of their rockets reusable; most notably the main booster stages, which are able to touch down under their own power, ready for reuse in, potentially, just a few days. Musk's avowed intent in setting up SpaceX

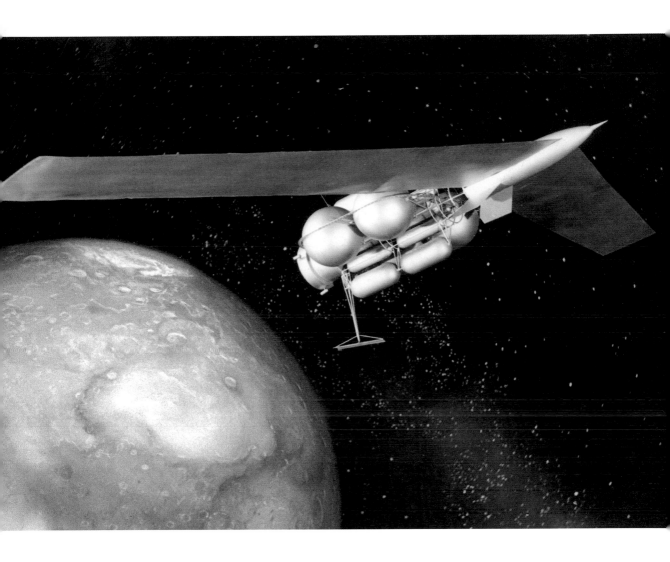

has been to develop the technology and capability to launch a crewed Martian expedition, and his plan to achieve this involves, like von Braun's, robust, reusable rockets capable of lifting large masses into Earth orbit at an affordable rate. To this end, SpaceX are pursuing an ambitious accelerated development programme for their Starship rocket and its successor, the Super Heavy rocket. These rockets are specifically intended to achieve multiple, cheap, rapid-cadence launches to get a lot of mass into Earth orbit, to allow orbital construction and fuelling of a spaceship that can reach Mars and touch

ABOVE Design concept for von Braun's Martian expedition spaceship; note the extensive fuel tanks.

OPPOSITE Diagram outlining orbital manoeuvres for rockets inserting to Mars's orbit.

down on the planet. What makes the parallels between von Braun's *Project Mars* story and the SpaceX project particularly striking is that in the former the explorers reach Mars and discover that it is ruled by a premier named Elon. Whether or not Elon Musk has ever read *Project Mars*, his ambitious Mars-shot programme is a direct descendant of the vision of von Braun.

ABOVE SpaceX founder and Martian colonization advocate Elon Musk, not currently premier of Mars.

LEFT SpaceX's Falcon Heavy rocket — a multiple-booster rocket designed to lift heavy loads into orbit — blasts off in early 2018.

PART 4
MEDICINE
& BIOLOGY

ELECTRA'S MAGIC RAYS:
ROENTGEN'S X-RAYS AND THE STORY THAT PREDICTED THEM

The decade spanning the turn of the twentieth century – from 1895 to 1905 – saw an extraordinary efflorescence of scientific discovery in the field of physics, revealing phenomena from radiation to sub-atomic particles to relativity.

The discovery of X-rays was the first and arguably the lynchpin of this crop of breakthroughs, with catalyzing effects for science, the public perception of science and the shape of science fiction to come. But it was also preceded – and perhaps prognosticated – by science fiction, for incredibly, this celebrated scientific achievement was predicted in astonishing detail by a fantastical tale penned three years earlier.

X THE UNKNOWN

X-ray is the name given to highly energetic electromagnetic radiation with wavelengths in the range of 0.01 to 10 nanometres, sitting between gamma rays and UV light in the electromagnetic spectrum. The X-ray was named by the German physicist Wilhelm Conrad Röntgen (commonly spelled Roentgen in English), after he discovered the phenomenon while experimenting with a Crookes tube. A Crookes tube, named after the Victorian English physicist William Crookes, is a sealed glass tube from which the air has been evacuated, with a cathode (negative electrical terminal) fitted inside one end of the tube. When connected to a power source, such a device was believed to emit from its cathode a beam or ray of unknown energy or particles, called a cathode ray. It is now known that a cathode ray is a stream of electrons, and that one phenomenon associated with

ABOVE Wilhelm Roentgen, the German physicist who first characterized X-rays, and who, in 1901, was awarded the very first Nobel prize for physics as a result.

such a stream is that, as energetic electrons decelerate (for instance, on ploughing into the glass-walled end of a Crookes tube), they sometimes shed their energy in the form of photons; that is, as electromagnetic radiation. This is known as *Bremsstrahlung*, or braking radiation, because it is emitted as the electron decelerates. This radiation may be sufficiently energetic to register in the X-ray region of the spectrum.

On the evening of 8 November 1895, Roentgen was experimenting with a Crookes tube sheathed in black card, in a darkened room. Whether by design or by accident, he had positioned on the far side of the room a sheet of paper coated with barium platinocyanide, a compound that fluoresces when energized with radiation. Roentgen noted that the paper glowed when he powered up the Crookes tube. He was confident that the cathode ray inside the tube could not penetrate the glass walls of the tube, and certainly could not pass through the sheath of card or the intervening atmosphere. Some other, unknown sort of energy must be emanating from the tube. "For the sake of brevity," he wrote in his account of the experiment, "I should like to use the term 'rays', and to distinguish them from others I shall use the name 'X-rays'." As in algebra, Roentgen made use of the letter X to designate an unknown factor.

What proved truly transformational was what Roentgen did next, as he systematically experimented with the X-rays to determine their properties and powers. Most remarkably, he discovered, X-rays are able to penetrate and pass through a wide range of substances. Roentgen was able to record the degree to which substances hinder the passage of X-rays by aiming his X-ray source at fluorescing screens and photographic plates and intervening test articles and substances, most notably his wife's hand bearing a wedding ring. "If one holds the hand between the discharge apparatus and the screen," he recorded, "one sees the darker shadows of the bones within the much fainter shadow picture of the hand itself." Roentgen had discovered a tool of unparalleled power for the examination of what had previously been hidden: a way to make flesh itself transparent.

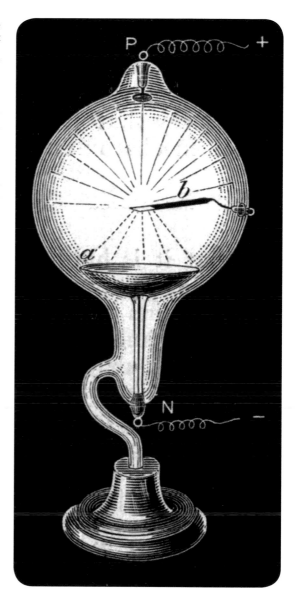

ABOVE An early example of a Crookes tube, a device that creates beams of electrons (known as cathode rays), which can, in turn, generate X-rays.

LEFT The basic set-up for the earliest X-ray photographs: a Crookes tube is aimed at a hand placed on top of a sealed photographic plate.

AS TRANSPARENT AS A JELLYFISH

Three years before Roentgen made his extraordinary discovery, a German physician named Ludwig Hopf, writing under the pen name Philander, had written a fairy tale on this exact same subject. In his story "Electra: A Physical Diagnostic Tale of the Twentieth Century", published in 1893 but written the year before, Hopf told the story of a young doctor around a hundred years in the future. Frustrated by a recalcitrant patient who refuses to let him perform a biopsy to confirm his diagnosis – "I would rather die than let myself be skewered alive!" – the doctor wanders into the countryside and muses aloud, "If only there were a means of making the human body as transparent as a jellyfish". To his amazement a woman appears in a blaze of light. She is Electra, the spirit of the twentieth century, and she puts into his hands a box that emits a strange radiation. Immediately a nearby tree

becomes, just as the doctor had wished, "as transparent as a jellyfish"; so does a frog examined with the aid of the box. The doctor rushes home and uses the device to confirm his diagnosis. The box is a simple electrical device and the doctor shares the details of its construction with the world, becoming a famous benefactor of humanity.

Philander/Hopf got one thing wrong: it would take much less than a hundred years for his fairy tale to come true. In fact, Roentgen presented his new discovery – and its application – to the Würzburg Physico-Medical Society in January 1896, and went on to win the inaugural Nobel Prize for physics in 1901, "in recognition of the extraordinary services he has rendered by the discovery of the remarkable rays".

Within weeks, Roentgen's discovery was being used to aid diagnosis. A front page article in the 26 January edition of the *New York Sun* remarked that,

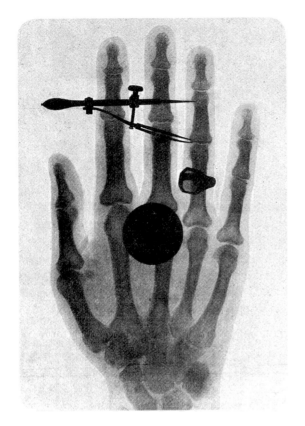

Never in the history of science has a great discovery received such prompt recognition and has been so quickly utilized in a practical way as the new photography which Professor Roentgen gave to the world only three weeks ago. Already it has been used successfully by European surgeons in locating bullets and other foreign substances in human hands, arms and legs and in diagnosing diseases of the bones in various parts of the body.

By 3 February that year the technique had been employed for the first time in North America, after an enterprising photographer in Dartmouth, New Hampshire, having read about the discovery, contacted his local physics department to suggest that if they had tubes that would generate X-rays, he would supply the photographic

ABOVE LEFT Roentgen's radiograph of his wife's hand, along with a protractor. Her ring shows up clearly as the gold parts are opaque to X-rays.

ABOVE RIGHT An early radiograph showing an American frog; X-ray photography offered an extraordinarily powerful new tool for scientists of all persuasions.

plates. A teenager named Eddie McCarthy, who had broken his arm ice skating, promptly became the first patient diagnosed by X-ray in the US. Before the year was out the Glasgow Royal Infirmary in Scotland had set up one of the first radiology departments in the world, achieving a number of firsts, including the first X-ray of a kidney stone, an X-ray showing a penny in the throat of a child; and an image of a frog's legs in motion.

The Sun.

THE WEATHER PREDICTION
For New York and Vicinity:
Generally fair; winds becoming westerly.

VOL. LXIII—NO. 148.   NEW YORK, SUNDAY, JANUARY 26, 1896—COPYRIGHT, 1896, BY THE SUN PRINTING AND PUBLISHING ASSOCIATION.—THIRTY-TWO PAGES.   PRICE FIVE CENTS.

FAITH IN THE ALLIANCE.

STRONG BELIEF THAT TURKEY AND RUSSIA HAVE A COMPACT.

THE NEW PHOTOGRAPHY.

REMARKABLE OUTCOME OF PROF. RÖNTGEN'S DISCOVERY.

ST. PAUL HARD AGROUND

A Race with Campania Ends on the Jersey Sands.

DID THE CUNARDER HIT, TOO?

Mails and Passengers Taken Off the Stranded Ship.

THE ST. PAUL AGROUND AND THE BREECHES BUOY CARRYING A LETTER ABOARD.

HOW THE ST. PAUL WENT ASTRAY IN THE FOG.

ABOVE Young Eddie McCarthy lays his broken arm on a photographic plate while a Crookes tube blasts it with X-rays to create the first radiograph in America.

OPPOSITE The front page of the *New York Sun* for 26 January 1896, bearing under the headline "The New Photography" an article about the new field of radiography.

STEAM-ENGINE-TIME

Is there any causal link behind the remarkable synchronicity between Hopf's story and Roentgen's discovery? There is no evidence that Roentgen was aware of the tale, and it is hard to see how it could have inspired his research. Rather, Roentgen's discovery appears to be a classic example of following up unexpected results without preconceptions. It is more feasible that there could have been some link between Hopf's tale and the rapid application of X-rays to medical science, except that the tales of Philander were not widely known at the time and not translated into English until relatively recently.

Perhaps the true connection between the two is that both story and discovery came about because it was their "time". A recurring phenomenon in the history of scientific discovery and technological invention is that breakthroughs are made almost simultaneously by multiple independent parties, as if it were their "time" to happen and so they somehow became available. The American philosopher and chronicler of anomalous events Charles Fort, pondering the way in which steam-engine technology had not leapt forward until the Victorian era, despite the principles being well known in ancient times, observed that "a [society] cannot find out the use of steam engines, until comes steam-engine-time". The phenomenon applies to many – perhaps the majority of – scientific breakthroughs, from calculus and the theory of evolution by natural selection to the light bulb and radio transmission. This train of thought was developed by sociologists of science such as the American Robert K. Merton, who pointed out that many discoveries are made in the context of the "adjacent possible": the way that prior advances make new ones possible in adjacent regions of "idea space". Thus, new ideas only become available when previous steps have been taken and the ground is prepared for them to flourish.

Speaking specifically of X-ray technology and the Electra story, medical historian and translator of Philander Paul Potter points out that Hopf was writing in the context of many contributory developments relating to germ theory and organ-based diagnosis, overtaking pre- and proto-scientific models such as humoral theory.

These meant that "Scientists were concentrating on finding ways of detecting internal changes before their patients became cadavers" – in other words, there was mounting pressure for a technique that would provide a window into the living body. This was the background against which Hopf penned his fairy tale: it was a fictional expression of medical wish fulfilment, imagining into being a power that was dearly wished for and eagerly sought. When Roentgen's rays made such a power available, it was seized upon by a profession that had been waiting for precisely this development.

UNLEASH THE RAYS

While sci-fi may not have played a direct role in the development of Roentgen's rays, his rays would have far-reaching consequences for the development of sci-fi. X-ray powers became a common feature of sci-fi. E. E. "Doc" Smith used the concept of "spy rays" repeatedly in his Lensman series, having coined it in his 1934 story, "Interplanetary". An alternative version was the "X-beam projector" of Gordon Giles's 1937 story, "Diamond Planetoid":

> As the pencil of soft X-rays, under the guidance of his skilled hand, probed into the twenty-foot planetoid, its reflections trembled ghostily in the milk-luminous chart … Osgood rotated the handle of the X-beam projector, the "X gun", with the expertness of a machine gunner …

Sci-fi rays did not stop at "X", however. Roentgen's discovery, which proved to be one of the most high-profile scientific breakthroughs of the age, helped to make rays a ubiquitous feature of public discourse about cutting-edge science, in similar fashion to the way that "quantum" is used today. Rays of every conceivable sort quickly became sci-fi tropes and then clichés, and every space opera featured ray guns galore.

OPPOSITE Ray guns quickly became characteristic – even clichéd – in science fiction, as demonstrated here on the cover of the January 1934 edition of pulp magazine *Amazing Stories.*

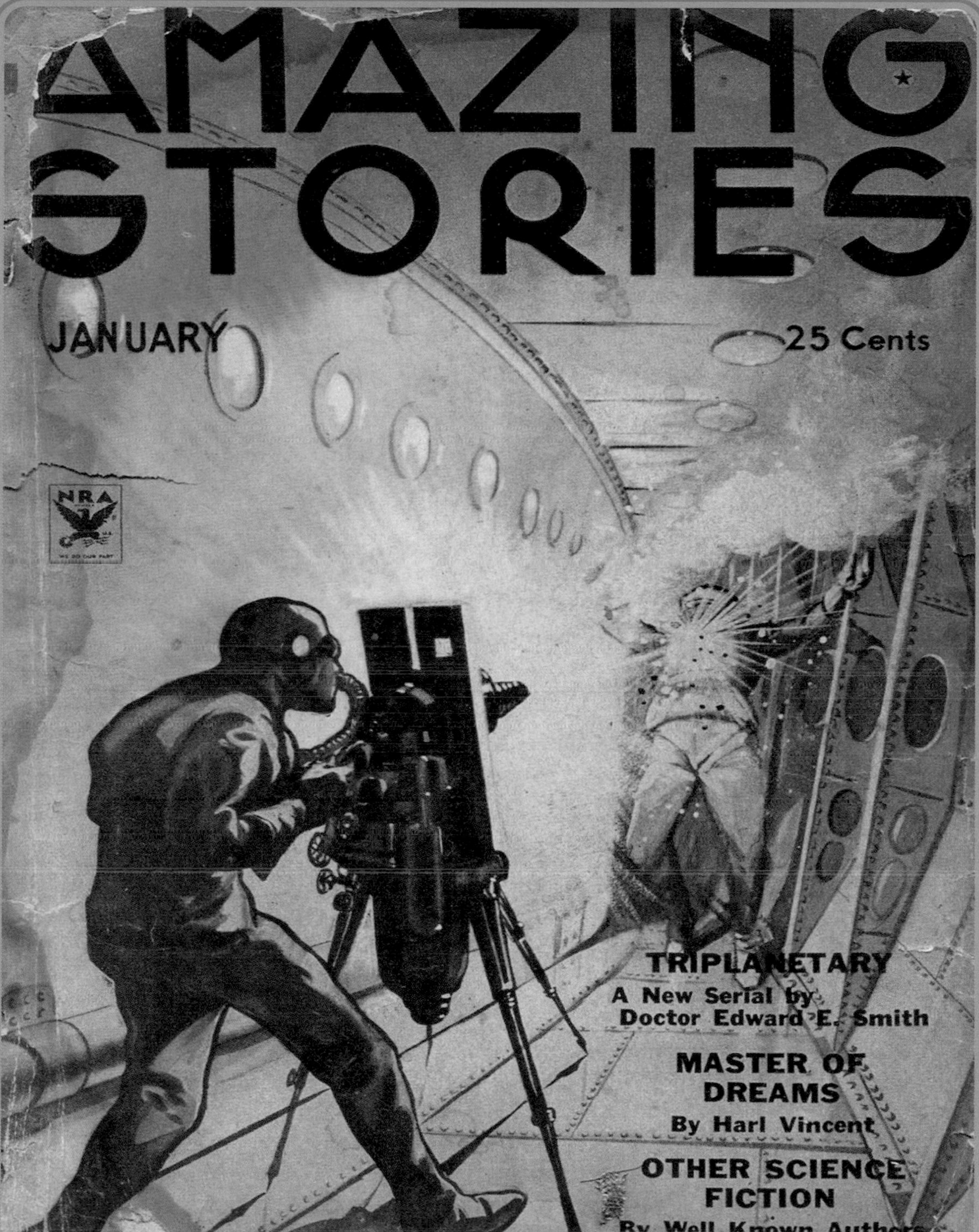

AMAZING STORIES

JANUARY

25 Cents

TRIPLANETARY
A New Serial by
Doctor Edward E. Smith

MASTER OF DREAMS
By Harl Vincent

OTHER SCIENCE FICTION
By Well Known Authors

13

ORGANISM ENGINEERING:
THE ISLAND OF DR. MOREAU, GERMLINE TINKERING AND INTERSPECIES CHIMERAS

In H. G. Wells's 1896 novel *The Island of Dr. Moreau*, a castaway comes to a mysterious South Pacific island inhabited by bizarre human-animal hybrids created by the titular mad scientist.

Like Frankenstein before him, Dr Moreau has become a touchstone for modern anxieties about hubristic science. These encompass the visceral unease caused by scientific encroachment on the ill-defined border regions between the natural and the artificial – or, perhaps more descriptively – unnatural. What links Dr Moreau's abominations with the Darwinian revolution in biology and modern aspects of biological science such as genetic engineering, germline tinkering, transgenic chimeras and radical transplant technology?

A MORBID ABERRATION

In Wells's book a shipwreck survivor, Prendick, is rescued but, suspected of cannibalism, is set adrift once more, only to be rescued again by those in charge of a strange island community. Prendick's first intimation that something is awry comes as he surveys the crew of the boat that picks him up, describing them as an "amazingly ugly gang ... there was something in their faces – I knew not what – that gave me a queer spasm of disgust". The leader of the crew explains to him: "This is a biological station – of a sort." What sort gradually emerges, as Prendick spies strange "bestial-looking creatures" in the jungle and notices animal features amongst the doctor's staff. Moreau, it transpires, was a celebrated scientist driven out of England when his

researches into animal vivisection were revealed. On this remote island he is pursuing those researches, using radical surgical modification and transplantation – without pain relief – to craft animals into humanoid creatures capable of bipedalism and speech. Moreau has attempted to impose crude forms of culture and "civilized behaviour" on his creations, but when a puma-person upon which he is experimenting escapes and kills him, the island is thrown into chaos and the beast-people revert to their animal ways. Prendick is the only human survivor, and eventually flees the island on a dinghy that has washed ashore bearing the remains of shipwrecked sailors. His last sight of the island reveals wolf-beasts and "the horrible nondescript of bear and bull" scavenging the rotting human remains: "Frantic horror succeeded my repulsion ..."

Still disturbing over a century after its publication, *The Island ...* received mixed reviews at the time. The London *Daily Telegraph* condemned it as "a morbid aberration of scientific curiosity", while Wells himself would later call it "an exercise in youthful blasphemy". Its themes echoed contemporary anxieties about vivisection, which had recently aroused a great furore in Britain (see below), and the implications of Darwinism, which had become the dominant paradigm in biology while engendering a

OPPOSITE Frontispiece to H. G. Wells's *The Island of Dr Moreau*, showing the protagonist Prendick encountering a trio of Moreau's abominable creations.

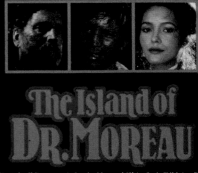

ABOVE Poster for the 1977 film adaptation of *The Island of Dr. Moreau,* including added love interest.

ABOVE Iconic image of Charles Darwin, whose theory revolutionized biology but was widely and often wildly misinterpreted.

RIGHT Birds and other creatures of the Galapagos, the island laboratory of evolution that Darwin made famous, perhaps inspiring Wells.

host of misunderstandings. Darwin's theory was widely understood to offer a progressive view of evolution as a process of improvement (viewed as a ladder), which in turn suggested that the process could be reversed, leading to degeneration. Anxieties about degeneration were further conflated with spurious racial science and poorly understood concepts of heredity. This was the context for Wells's beliefs about evolutionary theory.

Wells was a science journalist with just enough grounding in Darwin's theories to misunderstand them. He believed that the plasticity – capacity for change – of species over time, described by Darwin, implied a similar degree of plasticity in individual biology, so

that if science were sufficiently advanced, a creature might perhaps be extensively remodelled. Wells further believed that changes to the physical nature of an individual might produce cognitive and behavioural changes, so that, for instance, endowing an animal with the organs of human speech might allow it actually to speak. This was the central conceit of the novel and of Dr Moreau himself, whose dreadful experiments remodelled animals into para-human constructs, which accordingly gained para-human abilities. Taking his cue from Darwin's focus on the Galapagos as island laboratories of evolution, Wells sited his own evolutionary sandbox on a Pacific island.

CHIMERAS

Like its predecessor and inspiration, *Frankenstein*, Wells's novel reflected public perception that biology and medical science were advancing with exciting but also disturbing rapidity. Mary Shelley's novel, for example, was written in the context of increasingly accurate anatomical dissections, the birth of resuscitation medicine, advances in blood transfusion and, of course, the development of electrical science and galvanism in biology. In addition to Darwinism and vivisection flaps, Wells's novel came in the context of the rapid advance of cell theory (the recognition of the cell as the basic unit of biology), germ theory and microbiology, biochemistry and an increasingly detailed understanding of the processes involved in cell division and reproduction.

Chromosomes had been identified and observed undergoing division and replication, while a mysterious substance then labelled nuclein (the main component of which was later characterized as DNA) had been identified as a candidate for the medium of transmission of heritable characteristics. Indeed, heredity was one of the great preoccupations of the age, with widespread support for doctrines of eugenics (the improvement of humanity through control of heredity).

In 1896, however, many unknowns remained. DNA was regarded as a minor biochemical constituent of cells and the actual mechanism by which genetic information is encoded and transmitted would remain unclear for another 57 years. This makes it all the more striking that *The Island* ... so clearly articulates contemporary

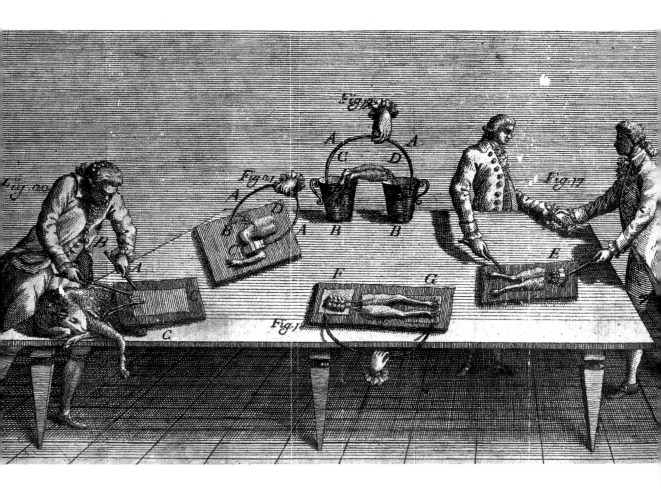

concerns about issues of biology and heredity, from transgenic chimeras to genetic engineering.

The Chimera or Chimaera was a fire-breathing monster from Greek mythology, a hybrid or mash-up of several animals. It was usually shown as having the body and head of a lion, but also the head of a goat coming from its back and, in place of a tail, the body

ABOVE Red-figure dish from fourth-century-BCE Ancient Greece, showing the mythical Chimera.

OPPOSITE Galvanic trials on frogs and other animals had helped inspire *Frankenstein*, and also set the template for scientists engaged in gory and macabre experiments.

and head of a snake. Although he does not use the term, Wells clearly implies that Moreau's creations are artificial chimeras, hybridized by surgery from parts of disparate organisms to create new, humanoid ones. But chimeras do exist in nature, and they are surprisingly common.

Biology co-opted the name chimera to describe a single organism made up of cells or parts from more than one genetically distinct individual (aka genotypes). Chimeras occur naturally; in animals, usually as the result of the fusion of embryos (for instance, of twins) at a very early stage in development, when they are just tiny balls of cells. The resulting offspring may have a few cells with a different genotype or whole tissues or organs. This can result in individuals with two blood types, or mosaics of skin or fur colour, or even, if the original two embryos were of different sexes, both male and female sex organs.

Chimerism might be much more common than is generally appreciated. In humans it usually happens when twin embryos are conceived but the zygotes (early stage embryos) merge so that only one baby is born. According to chimerism expert Charles Boklage, a developmental biologist at the Brody School of Medicine at East Carolina University, "about one in eight of everybody walking around is a twin who was born single", and thus has a good chance that they may incorporate cells from their original twin. Boklage points

OPPOSITE A genetically engineered mouse engaged in a navigation trial. Would it be ethical to make similar changes to a human being?

BELOW With microscopic precision, cells are added to an embryo to create a transgenic organism.

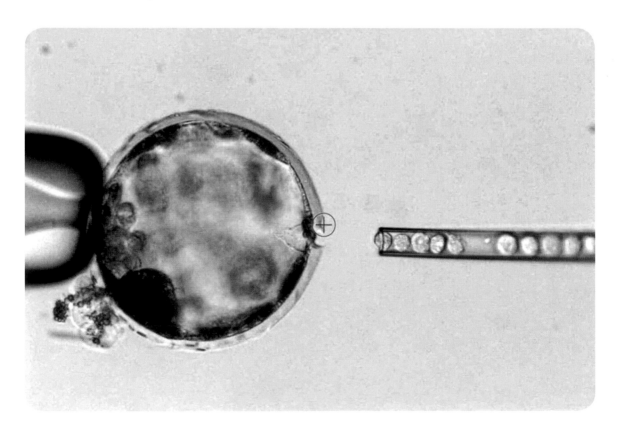

out that naturally occurring human chimeras raise many interesting and potentially disturbing questions about the nature of self and identity. The possibility that "a substantial fraction of us are made of cells from what might have been two people", he reflects, "has a really big 'yuck factor' for most people ... Are there ways that these people are different that we haven't seen?" Boklage even poses the question, "Philosophically, how many souls does a chimera have?"

But chimeras are more than just a biological curiosity. Creation of chimeras is one of the main research avenues being pursued to rectify chronic problems in organ transplantation, such as shortage of available organs and the problem of rejection. This is the premise behind the creation of transgenic animals – organisms with the genomes of more than one species present in at least some of their cells. The goal is to insert into the genome of an animal host human genes that determine whether or not an immune response will be triggered. For instance, if the cells in the kidneys of a pig can be made to express human genes for certain cell-surface molecules, those kidneys might be made to appear less foreign to the immune system of a human host. Such a pig would be transgenic. In the future, perhaps, it might even be possible to use the intended human host's own cells or genome as the seed or blueprint for these organs, so that perfectly compatible, human organs could be farmed inside chimeric animal hosts.

Schemes such as these arouse strong controversy and, often, instinctive revulsion of the very sort that permeates Wells's novel, with its "spasms of disgust". The creation of transgenic animals raises ethical concerns and particularly troubles religious observers, although the "yuck factor" involved in such science is much more widely applicable, appealing to ill-defined and poorly articulated concepts of the unnatural.

BRAVE NEW WORLDS

Similarly controversial is the science of genetic engineering, particularly the branch that involves germline alterations. These are changes to an individual's genome that can be passed on through

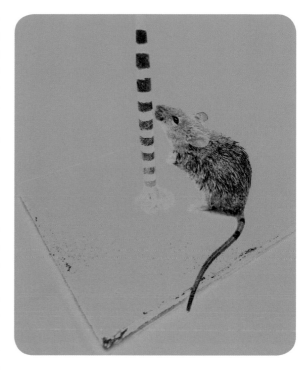

his or her gametes (reproductive cells), and hence are heritable by the next generation. Profound ethical concerns about the potential consequences or misuse of such technology have led to a global moratorium on germline genetic engineering. But in 2018, a Chinese scientist caused a storm of controversy and outrage when he announced that he had genetically engineered pre-implantation embryos that had been carried to term. Ostensibly, the children had been engineered to be more resistant to HIV, which seemed like a potentially spurious rationale for defying international norms, but it transpired that other modifications had also supposedly been made. The claims have not been independently verified and there is some degree of scepticism about them, because genetic engineering is notoriously hard to achieve *in vivo*. Nonetheless, this technology is clearly imminent if not already extant, so that genetic engineering may soon move out of the preserve of science fiction, where it has been one of the primary themes of the last 40–50 years.

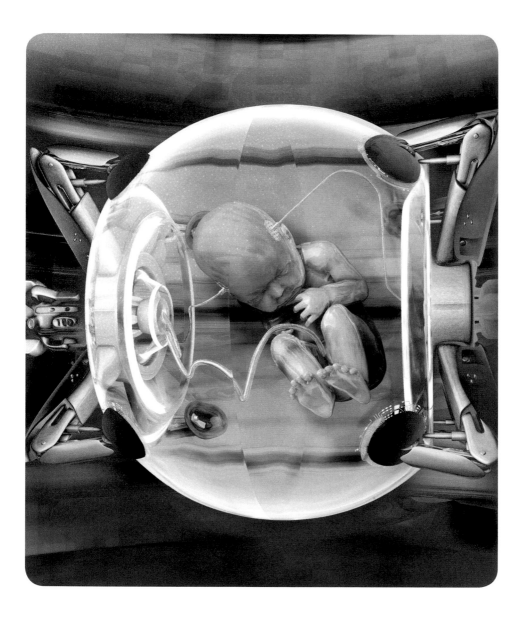

Sci-fi treatments of genetic engineering are too numerous to survey here, but the touchstones for this field of technology include Dr Moreau, alongside Frankenstein and Aldous Huxley's 1932 novel *Brave New World*. Intriguingly, Huxley's novel draws its title from one of the major influences on Wells's *The Island …*, the Shakespeare play *The Tempest*, in which shipwrecked sailors discover a renegade intellect exiled to pursue his researches in isolation, transgressing against nature and unleashing dangerous semi-bestial agents. Indeed, it is from *The Tempest* that we derive the terminology of nature versus nurture, when Prospero describes his bestial servant Caliban as "A devil, a born devil, on whose nature, Nurture can never stick".

In Huxley's *Brave New World* the society of the future proves to be not only biologically determined, in the sense

that socio-economic status is determined exclusively by biological make-up, but this determination is artificially controlled. Embryos are gestated in machines and biochemically treated to determine their characteristics and intellectual abilities. Writing before the genetic revolution engendered by the 1953 discovery of the secret of DNA, Huxley did not have access to concepts of genetic engineering. Consequently, he does not use this terminology, but the themes are clearly applicable and his dystopian future is widely presented as the likely outcome of the slippery slope onto which society will be tipped once germline tinkering is allowed.

ABOVE Some surprisingly well-groomed beast-men, in a scene from the 1971 movie adaptation of *The Island of Dr. Moreau*.

OPPOSITE A dystopian view of the future of human reproduction, as envisaged in *Brave New World*.

H. G. Wells's *The Island of Dr. Moreau* has also been viewed as an exploration of the ramifications of unfettered genetic experimentation. The biology actually cited in the book does not stand up, as the experience of transplant recipients testifies, because the immune systems of organisms do not allow them readily to graft or bond with foreign tissues and cells (chimeras are interesting exceptions, presumably viable because the relevant immune recognition systems have habituated to the "foreign" cells since the earliest stages of development). But if the precise mechanics employed to facilitate the central conceit are elided, or substituted for something more plausible in light of current knowledge, it is possible to imagine some form of transgenic hybridization akin to Moreau's awful experiments. Genetic engineering to incorporate genes and attributes from other organisms is a common proposal for a far-future technology to enable human conquest of extreme environments such as outer space, the deep ocean or the climate-changed world of the future. To give just one example, Lois McMaster Bujold's 1988 novel *Falling Free* concerns "Quaddies": people who are genetically modified to develop an extra pair of arms instead of legs, equipping them for the zero-gravity environment in which they live. In the corporate culture of the novel, the Quaddies are considered not people but "post-fetal experimental tissue cultures", and a culture that privileges values of corporate exploitation is imposed upon them by their creators, just as Moreau attempts to impose a culture on his creations.

ABOVE Frances Power Cobbe (1822–1904), pioneer of the animal rights movement and anti-vivisection campaigner; she would have taken a dim view of Dr Moreau's methods.

OPPOSITE Posters in the window of the British Union for the Abolition of Vivisection in 1936.

EXTREME TRANSPLANTS

Might future individuals and societies use advanced genetic engineering to achieve Moreau-like ends? Already today some individuals choose to undergo radical cosmetic and body modification surgery to look like human-animal hybrids. But an even more startling iteration of Wells's fever dream of cut-up biology lurks around the near-future corner. Wells wrote *The Island* ... partly in response to contemporary revulsion for the practice of vivisection, about which a moral panic had been stoked by the publication of the first English guide to the subject, the 1873 *Handbook for the Physiological Laboratory*. Vivisection was lambasted as a cruel import from the Continent, where it was seen as an important tool in the rapidly advancing field of experimental medical science. Wells may have drawn directly on the testimony of Emmanuel Klein, an Austrian vivisector working in London who gave evidence to a Royal Commission in 1875, insisting that he was utterly indifferent to animal pain and rarely used anaesthetic in his vivisections. Moreau perpetrates his outrages without anaesthetic, in an operating theatre that the

Beast People come to know as the "House of Pain".

But vivisection in many ways was the opposite of the technology Wells envisaged in his novel. While medical scientists had become increasingly adept at operating on living organisms, their powers were limited to destruction, not construction. Although a type of human tissue grafting – blood transfusion – was achieved in the early nineteenth century, it would not be until the 1950s that successful human organ transplants were achieved. Baleful sci-fi tropes such as *Frankenstein* were widely employed in discussing such advances; for instance, in 1967 the recipient of the first human heart transplant likened himself to Frankenstein's monster. The technology, however, has continued to advance, so that today most of the major organs can be transplanted, as can body parts such as hands and even faces. The key advance has been immunosuppressant drug therapy, which allows transplant recipients to restrain their immune response so that organs are not rejected.

Such advances have made thinkable scenarios that were once relegated to science fiction, in particular one reminiscent of both Frankenstein and Dr Moreau: the head transplant (or perhaps more properly, the body transplant). The spectre of this extraordinary and ethically disturbing proposition was raised by Moreau-like experiments, initially on dogs in the 1950s and later on a monkey in 1970. Soviet transplant pioneer Vladimir Demikhov showed that the head and forelimb of one dog could be surgically grafted onto the body of another and survive, at least for a few days. Inspired by this research, in 1970 American neurosurgeon Robert White attached the severed head of one rhesus monkey to the body of another. The head regained consciousness and was able to blink and look around, but the monkey was paralysed and died after eight days because of tissue rejection.

While these macabre experiments have attracted the kind of opprobrium that drove the fictional Moreau to his island retreat, they have apparently not discouraged the controversial Italian neurosurgeon, Sergio Canavero, who announced in 2016 that he would shortly perform the first human head transplant. Canavero claimed to have learned how to overcome the previously insurmountable problem of reconnecting

OPPOSITE Boris Karloff as Frankenstein's monster, the touchstone for a gamut of scientific, technological and medical anxieties.

BELOW The first heart transplant in progress; human transplant medicine met with moral objections and dark prognostications when it first became a reality.

a severed spinal column, with a combination of organic glue and Frankenstein-inspired electrical stimulation; he said that he and his Chinese collaborator Xiaoping Ren had successfully demonstrated the technique on mice, monkeys and human cadavers. He further claimed that a disabled Russian man had volunteered to have his head transplanted onto the body of a brain-dead donor. Alongside widespread scepticism that the surgery had any chance of success, there were accusations that the whole affair was a publicity stunt and it petered out.

Nonetheless, the prospect of this surgery raised intriguing philosophical and psychological questions that equally apply to Moreau's monsters. What kind of consciousness might result from extreme body hybrids of this sort, especially given the evidence that embodiment plays a crucial role in cognition and consciousness? In metaphysical terms, would a composite creature have many souls or just one, and if the latter, to which contributor would it belong?

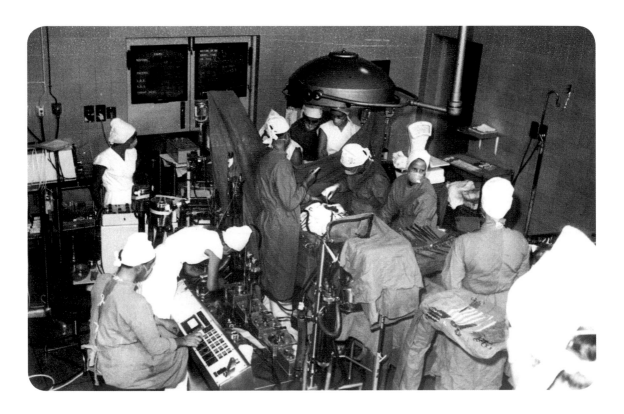

14

PROZAC NATION:
HUXLEY'S SOMA AND ANTIDEPRESSANTS

Psychotropic drug consumption on an industrial scale, the mood and behaviour of a nation controlled by pharmaceuticals, a population rendered robotic and compliant by state-sanctioned pill popping.

This was the chilling dystopia foreseen by British writer Aldous Huxley in his 1932 novel *Brave New World*, but according to many concerned observers it was also not far off the reality of late twentieth-century America after the release and massive success of the new antidepressant Prozac. How accurately did Huxley predict the advent of pharmaceutical mood regulation on a massive scale, and should modern society attend more closely to his warnings across nearly a century?

PHARMACOLOGICAL TOTALITARIANISM

In the future society of *Brave New World*, the working masses are kept psychologically sated by liberal application of a drug called "soma", which might be viewed either as a universal panacea, or, as described by writer and historian of sci-fi Robert Silverberg, a "contemptible anodyne". "If ever by some unlucky chance such a crevice of time should yawn in the solid substance of their distractions," Huxley wrote, "there is always soma, delicious soma, half a gramme for a half-holiday, a gramme for a weekend, two grammes for a trip to the gorgeous East, three for a dark eternity on the moon; returning whence they find themselves on the

LEFT English novelist and essayist Aldous Huxley, scion of a family prominent in the history of science, and biology in particular.

other side of the crevice, safe on the solid ground of daily labour and distraction ..."

Soma is a Latin word, meaning "body", but it was also the name of a mysterious drug or medicine mentioned in the ancient Indian religious texts known as the Vedas. Enormous quantities of ink have been spilled in the attempt to identify this ancient and possibly prehistoric iteration of soma, which is usually understood to be some sort of psychedelic mushroom or plant preparation, probably either psilocybin, cannabis, opium or some combination of these. Huxley was a pioneer of psychedelics who would go on to proselytize for their use, but in creating soma he had much darker and more insidious motives in mind. As he explained to an audience at the University of California, Berkeley, in 1962, his "hypothetical drug", which was "simultaneously a stimulant, a narcotic, and an hallucinogen" opened up a new zone of control for the totalitarian state:

> ... an enormous area in which the ultimate revolution could function very well indeed, an area in which a great deal of control could be achieved not through terror, but by making life seem much more enjoyable than it normally does. Enjoyable to the point where ... human beings come to love a state of things which by any reasonable and decent human standard they ought not to love, and this I think is perfectly possible.

In a survey of drug themes in science fiction, which he compiled for the US National Institute on Drug Abuse, Silverberg pointed to other treatments of drugs in sci-fi similar to Huxley's. In James E. Gunn's *The Joy Makers*, published in 1961 but written half a decade earlier, mandatory use of euphorics maintains a repressive government in office, as in L. P. Hartley's dystopian *Facial Justice* (1960). George Lucas's 1971 film *THX 1138* has a similar scenario, with mind-controlling drugs used to ensure that a repressed population follows orders and perseveres at dangerous and demanding work. In fact, Lucas's film is clearly heavily inspired by

ABOVE Leonard Nimoy as Mustapha Mond, Resident World Controller of Western Europe, in the 1998 TV movie adaptation of *Brave New World*.

Brave New World. In Huxley's novel the satiating effects of soma are backed up by the broadcast of erotically gratifying "feelies" (virtual-reality entertainments), so that the population is provided with ersatz bread *and* circuses; in *THX 1138* the workers take pills and use a masturbation device to gratify their sexual urges.

FROM TANK CHOCOLATE TO PROZAC

In the decades after Huxley's novel was published, rapid advances in psychotropic and psychotherapeutic drug research brought into ever-sharper focus the future he had sketched out. Large-scale mind modification through chemistry received a massive boost from the exigencies of war, as both sides in the Second World War sought to boost the alertness and endurance of their militaries. Particular favourites were amphetamines, which acquired a variety of nicknames, including "go pills", "pep pills" and "tank chocolate" (aka *Panzerschokolade*, scoffed like candy by German tank crews). Among the bigwigs boosting pill popping was the British general Bernard Montgomery, who was converted to the chemical cause after observing the results of field trials in which a drugged infantry squad won a race against a "straight" one, the tanked-up Tommies deemed to have displayed

ABOVE Wartime advertisement for benzedrine, somewhat coyly extolling the virtues of the amphetamine "for relief of nasal congestion".

OPPOSITE In George Lucas's 1971 film *THX 1138*, a repressive state uses drugs to control and enslave its population.

impressive levels of "snap and zest". Accordingly, large quantities of amphetamine pills were procured for British troops in the build-up to the Battle of El Alamein in 1942. By the following year the Americans switched on to the potential application of amphetamines, and benzedrine sulphate tablets, or "Bennies", were produced in huge numbers. American general Dwight Eisenhower requested half a million packs for his troops, and abuse of such pills was linked to the commission of battlefield atrocities in the Pacific theatre.

After the war such drugs had a huge impact in psychiatric medicine, and although medications such as thorazine were often stigmatized for "zombifying" patients, they revolutionized the treatment of previously intractable conditions such as schizophrenia. But pharmaceutical manufacturers in search of new markets pushed mood- and behaviour-altering pills in ways that were heavily criticized for medicalizing behaviour and cognition that seemed to many to fall within the range of "normal". So, for instance, children were fed amphetamines under the brand name Obetrol to fight obesity, while tranquilizers were marketed as housewives' pick-me-ups and amphetamine-like Ritalin were prescribed to millions of children.

By 1962, when he addressed an audience at Berkeley, Huxley pointed out that while his hypothetical drug soma had seemed unlikely when he invented it, "if you applied several different substances you could get almost all these results even now ..." This may not have been entirely accurate, but around this time a new type of drug was coming into clinical use: antidepressants. Although the early classes of antidepressant had quite severe side effects, they proved highly effective,

saving lives and providing hope to sufferers of acute depression. It was the successor to these initial classes that would ape the claims made for soma. In 1987 a new antidepressant called fluoxetine, belonging to a class of drug known as selective serotonin reuptake inhibitors (SSRIs), was brought to market under the name Prozac. It has proved to be highly effective, with dramatically fewer side effects than its predecessors, and a combination of word-of-mouth about its effectiveness, clinical enthusiasm, and a massive marketing campaign by its makers would quickly make it the bestselling antidepressant of all time.

Since then Prozac and related SSRIs have been targeted at an ever-increasing range of users, and in addition to even quite mild depression, it is now commonly prescribed for bulimia, anxiety disorders and some forms of behavioural disorder in children. It was even licensed in the US for the treatment of severe premenstrual syndrome. From 1990 to 2011, according to the Centers for Disease Control and Prevention's National Center for Health Statistics, antidepressant use in the United States rose by nearly 400 per cent. In America, more than 10 per cent of people over the age of 12 take antidepressants and they are the second most prescribed class of drug in medicine.

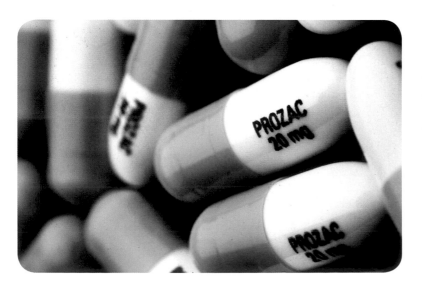

LEFT Capsules of the antidepressant fluoxetine, trade name Prozac.

OPPOSITE Still from the 1980 film adaptation of *Brave New World*, starring Keir Dullea (of *2001: A Space Odyssey* fame) and Marcia Strassman.

PHARMACOLOGICAL CALVINISM

The runaway popularity of Prozac brought a backlash. It was claimed that the drug acted to smooth out peaks and troughs in normal mood, producing "Stepford Wives"-style equanimity; that it suppressed creativity and left users feeling numbed and disengaged; that it was a tool for enforcing conformity and mediocrity; even that its massive overprescription amounted to a form of state-sanctioned thought control. While many of these claims may have shaky foundations at best (for example, it is also claimed that Prozac can enhance creativity by relieving the burden of depression), it is recognized that

a potential side effect of SSRI use, affecting up to 20 per cent of users, is "flattening of affect": a kind of emotional flattening or dullness.

Even before Prozac had come on the market there were arguments in the psychiatric community that drugs were being overprescribed, and that medicating mood or troublesome cognition is a vice rather than a virtue. In the 1970s, Dr Gerald Klerman used the term "Pharmacological Calvinism" to describe the view that taking drugs to control mood is somehow cheating or inauthentic. Huxley had anticipated precisely this view in *Brave New World*:

There's always soma to calm your anger, to reconcile you to your enemies, to make you patient and long-suffering. In the past you could only accomplish these things by making a great effort and after a year of hard moral training. Now, you swallow two or three half-gramme tablets, and there you are. Anybody can be virtuous now. You can carry at least half your morality about in a bottle.

MAX HED-ROOM

Huxley's soma puts a pharmaceutical spin on a philosophical conundrum best known today through the work of American philosopher Robert Nozick, and his "Experience Machine" thought experiment. Nozick wished to interrogate some of the assumptions underlying how happiness or contentment (aka well-being) is defined and measured. The dominant approach in this area has been informed by the branch of philosophy known as utilitarianism, which is based on the maxim that "Pleasure and only pleasure is good." This is a position known as philosophical hedonism. What determines whether an experience contributes to well-being, according to the utilitarians' philosophical hedonism, is its pleasurability. Well-being is maximized when pleasure is maximized and pain is minimized.

Nozick imagined an "Experience Machine" into which people could plug, which would give them the experience of maximizing their pleasure and minimizing their pain, but only as a virtual reality illusion (see page 218 for more on this, and its parallels in science fiction, such as in the film *The Matrix*). But he might just as well have cited Huxley's soma as a pharmaceutical equivalent: a drug that produces maximal hedonism. Nozick said that people offered the chance to plug into the Experience Machine will reject it in favour of real life, partly because they value authentic experience and identity, but more profoundly because hedonism is not the true path to well-being. In Huxley's dark vision, however, people gladly opt in to his medically mediated equivalent of the Experience

ABOVE Robert Nozick, the American philosopher who proposed the Experience Machine thought experiment.

OPPOSITE Jack Nicholson in *One Flew Over the Cuckoo's Nest* (1975), the touchstone for the anti-psychiatry movement that preached a kind of "Pharmacological Calvinism".

Machine. As prescriptions for psychotropic drugs continue to rise, and as new, more effective forms of such drugs come to market, the questions posed by Huxley's soma grow more urgent. Will society resist the lure of the pharmaceutical panacea? Will we also cop out by opting in?

15

BIONIC PEOPLE:
FROM *THE SIX MILLION DOLLAR MAN* TO THOUGHT-CONTROLLED PROSTHETICS

An experimental aircraft out of control ... a terrible crash ... a solemn voice-over intones the immortal lines: "Gentlemen, we can rebuild him. We have the technology."

For a generation of children – and plenty of adults – who watched television in the 1970s, the opening sequence of *The Six Million Dollar Man* is familiar to the point of cliché. The unfortunate pilot in the crash was astronaut and test pilot Steve Austin (played by Lee Majors), and the technology referred to was bionics. Austin was transformed into a very expensive cyborg, whose artificial body parts (including both legs, one arm and one eye, all drawing juice from an unspecified atomic power plant) gave him superpowers, making him "Better ... stronger ... faster". The series started off as a television movie, itself based on the 1972 book *Cyborg*, by pilot, aerospace expert and author Martin Caidin. What were the antecedents of the Bionic Man, and how did the massive popularity of the series impact the real-world science of bionics?

CYBERBUGS

The terms "bionic" and "cyborg" derive from the science of cybernetics. Bionic is a contraction of "biological electronics", and cyborg a contraction of "cybernetic organism". Defined in 1948 by the father of the discipline, Norbert Wiener, cybernetics is "the scientific study of control and communication in the animal and the machine". It studies how systems are controlled and regulated by means of feedback from their environments.

This led cyberneticists to investigate, for example, how animal limb movements and locomotion are constrained, enabled and regulated by the way they interact with the environment (such as the way that a leg hinges off a hip and presses on the ground), and also how to create simple electromechanical systems that mimic or replicate animal behaviours.

Wiener, for example, helped create a cybernetic "bug" or "moth", named Palomilla: a three-wheeled cart equipped with a motor and steering mechanism linked to a pair of simple photoreceptors. Depending on how it was wired up, the cart would move either towards or away from a light source, like a moth or a bug exhibiting phototropism. Tweaking the signal amplification in the circuit could produce naturalistic behaviours such as tremors, similar to those seen in Parkinson's disease.

In studying how electronic circuits and nervous systems work to produce and control movement, cybernetics offered clear applications for fields such as robotics and prosthetics. At the intersection of these two fields is the world of bionics and cyborgs: fertile ground for science fiction.

WE CAN REBUILD HIM

This rich territory predated Wiener and cybernetics, however. The concept of replacement body parts is at least as old as recorded history. Ancient prostheses include a big toe discovered fixed to an ancient Egyptian mummy, while the third-century-BCE Roman general Marcus Sergius supposedly had an iron prosthetic

ABOVE Norbert Wiener, father of cybernetics, with one of his cybernetic creations.

hand that was so effective he was able to continue his career for many years. Replacing an entire person with prosthetic or artificial parts produced mythical constructs such as the bronze giant Talos.

The obvious touchstone in early sci-fi for this kind of technology is *Frankenstein*, published in 1818. Where Steve Austin was rebuilt with electromechanical parts, Mary Shelley's protagonist uses entirely organic parts, but like the Bionic Man's creators, he must have had an impressive command of cybernetics to put together such a complex system. One of the direct progenitors of Frankenstein's monster was the Golem: in medieval Jewish legend an inorganic simulacrum of a human brought to life through magic to become an unstoppable automaton, which escapes the control of its creators and

runs amok. It is interesting to note that Norbert Wiener explicitly invoked the Golem in warning of the perilous potential of untrammelled cybernetic advances – in other words, of the dangers of developing machines sophisticated enough to take over, which once switched on, cannot be switched off.

Direct precursors of Steve Austin can be traced back to the nineteenth century. In Edgar Allan Poe's satirical 1839 story "The Man That Was Used Up", a handsome soldier proves to have been almost entirely replaced with prosthetic parts. Poe's "Used Up Man" is perhaps the earliest iteration of figures such as Darth Vader or Saw Gerrera of the *Star Wars* universe, cyborgs who have sacrificed a piece of their humanity with every organic body part that is bionically replaced. In the 1917 story "Blood and Iron" by Perley Poore Sheehan and Robert H. Davis, later made into a play, a cyborg is created for the German kaiser by rebuilding a mutilated soldier with mechanical parts; like the Golem or Frankenstein's monster, he runs amok and ends up killing the kaiser.

POWERFUL STORY BY **JOSEPH CONRAD**

SOUTHAMPTON STREET

THE STRAND MAGAZINE

"BLOOD and IRON"

THE FATE OF THE KAISER

The Most Grim and Terrible Dramatic Scene ·Ever Written·

Nº 322 VOL 54

Daimler
DU 1916

8 d. net.

OCTOBER, 1917.

Published monthly by GEORGE NEWNES, Ltd , 8 to 11, Southampton Street, Strand, London, England.

An even more direct predecessor of the Bionic Man is the protagonist of Raymond Z. Gallun's 1935 story "Mind Over Matter": a test pilot who is nearly killed in a crash but whose brain is rescued and housed in an android body.

ABOVE A rabbi puts the finishing touches to the Golem of Prague, a magically animated automaton, in the pioneering 1915 horror movie *Der Golem*.

RIGHT An iron prosthesis from the sixteenth century, believed to have belonged to a German knight.

OPPOSITE Cover of the *Strand Magazine* from October 1917, featuring "Blood and Iron: The Fate of the Kaiser" — the "most grim and terrible dramatic scene ever written".

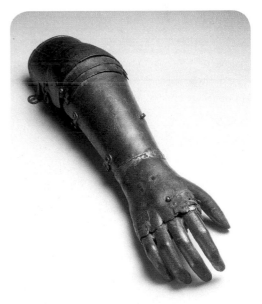

IN SEARCH OF SKYWALKER'S HAND

Perhaps Steve Austin would have been better served if, like Gallun's hero, his whole body had been replaced; for although the series never addressed the issue, his bionic exploits would have instantly destroyed the flesh-and-blood parts of his body. In the show, Austin is able to lift heavy loads with his bionic arm and jump from great heights, landing safely on his bionic legs. But even if his bionic limbs were capable, through advanced engineering and materials science, of exerting tremendous forces and bearing huge stress, the surviving parts of his body – especially the skeleton to which his bionic limbs were anchored – would have snapped or shattered immediately. Obviously the series' makers were more interested in entertainment than scientific plausibility or biomechanical realism, but it is illuminating to consider such challenges in light of the attempt to achieve genuine bionic prostheses.

The Six Million Dollar Man set an extremely high bar for public perception of what bionics might look like. The show made little effort to make Austin's prostheses look any different from the rest of his body, meaning that they were not only super-powered but also totally life-like. Perhaps the most influential representation of a bionic prosthesis in pop culture is the bionic hand seen being fitted to Luke Skywalker at the end of the 1980

OPPOSITE Recurrent foes of Doctor Who in the British sci-fi TV series, the Cybermen are cyborgs who have become more machine than human.

BELOW Lee Majors as Steve Austin in *The Six Million Dollar Man*, complete with remarkably life-like bionic body parts.

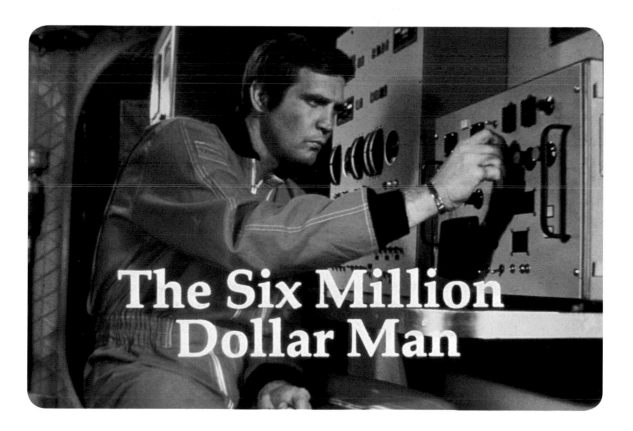

Star Wars film *The Empire Strikes Back*. The artificial skin of the prosthetic forearm is folded back to reveal its internal electromechanical workings, and the prick test performed by the robot surgeon explicitly demonstrates that the prosthesis has life-like haptic (sense of touch) feedback, replicated reflexes, and a seamlessly integrated interface between its electronics and Luke's own neuromuscular system.

ABOVE Luke Skywalker battles the cyborg Darth Vader; note Luke's right (gloved) hand, which is bionic.

Achieving this level of technological sophistication in real life has faced several major challenges: miniaturization, power supply, neuromuscular connection and stump attachment. Some of these have now been largely overcome; electric motors can now be extremely small, and high-level computing power is available to guide the action of the limb. So the mechanized prostheses themselves have more or less arrived, but much tougher problems remain on two other fronts. Attaching a bionic limb to a stump remains problematic; temporary attachments, which are the norm, cause long-term damage and limit the extent to which prosthesis and user can be connected, which in turn impacts the other main problem area: communication. How can messages be passed from the user to the prosthesis to control movement, and how can they be transferred the other way to enable sensory feedback from the prosthesis, including the sense known as proprioception (knowing where your body is)?

Currently the most widely available attempt to meet these challenges is with detachable prosthesis that use myoelectric sensors for control. Myoelectricity is muscle-generated electricity; electric pulses generated by the muscles remaining in the user's stump, which respond when the user tries to "move" the missing limb. Sensors attached to the stump can pick up these electrical signals and transfer them to the prosthesis, where they can be used to control motors. But this technology only really works for arm and hand prostheses, and it is one-way only; no information passes back from the prosthesis.

BELOW Jesse Sullivan and Claudia Mitchell, two real-life bionic people, high five with their neuromuscular-input-controlled prosthetic arms in 2006.

LEFT "Bionic artist" Viktoria Modesta, a Latvian pop and performance artist and model, who challenges conventional aesthetic preconceptions.

OPPOSITE Johnny Matheny demonstrates his truly thought-controlled prosthetic arm. Note the myoelectric sensors/controllers on his upper arm.

PLUGGING IN

A more ambitious system under development by the company Cambridge Bio-Augmentation Systems (CBAS) involves a "port" on the end of the stump that is hardwired to the nervous system of the body, into which is plugged an advanced prosthesis. CBAS co-founder Oliver Armitage describes it as "a USB port for the body". The trickiest aspect is that the port is surgically attached to a stump, anchored to the bone. Such an attachment offers an immediate boost to proprioception because the prosthesis then becomes an extension of an existing skeletal element. But implantation surgery is risky, because it means leaving an artificial implant protruding from a break in the skin, massively raising the chances of infection. The intention is to use new biomaterials and techniques to encourage the user's skin to grow onto the base of the port to create a seamless connection, and for artificial sensors in the prosthesis to be able to feed into the user's sensory nerves to provide some level of sensory feedback. This technology is already available in cochlear and optic implants – bionic ears and eyes.

Perhaps the most advanced system currently being tested combines aspects of both approaches. The Revolutionizing Prosthetics programme at Johns Hopkins Applied Physics Lab in the US has developed a sophisticated bionic arm that is controlled by a combination of myoelectric and direct brain stimulus, with implants that pick up nervous impulses from the user's muscles, peripheral nervous system and motor and sensory cortices (parts of the brain), and send back the other way feedback from artificial sensors. In December 2017 Johnny Matheny, a Florida man who lost an arm to cancer, was fitted with the arm for testing. News coverage of the research extensively referenced *The Six Million Dollar Man* and *Star Wars*.

PART 5
COMMUNICATIONS

16

VIDEOPHONES:
FROM TELEPHOTS AND PICTUREPHONES TO SKYPE AND FACETIME

A scientist named Ralph wishes to speak to a colleague. "Stepping to the Telephot on the side of the wall, he pressed a group of buttons and in a few minutes the faceplate of the Telephot became luminous, revealing the face of a clean-shaven man about thirty, a pleasant but serious face. As soon as he recognized the face of Ralph in his own Telephot, he smiled and said, 'Hello, Ralph.'"

Change "Telephot" to Skype or Facetime or WhatsApp or one of many other video-calling apps, and this passage could describe a mundane occurrence in today's world. But easily accessible video calling only became available for the mass market as recently as 2006, while the story from which this passage is extracted dates back to 1911.

THE FUTURE WILL BE TELEVISED

The Telephot is one of several predicted future technologies that make the story *Ralph 124C 41+* one of the most cited examples of prescient science-fiction tales, and its author Hugo Gernsback a poster boy for sci-fi's power of prognostication. Gernsback was a keen student of science and technology, a follower of trends and a tireless advocate for the inspirational power of science fiction (see Chapter 4). He would go on to become a pivotal figure in the history of sci-fi (one of the field's leading awards, the Hugo, is named after him), though less well known as an author in his own right. The story of Ralph, initially published in the magazine *Modern Electrics* in instalments that were later combined to produce the 1925 novel *Ralph 124C 41+: A Romance of the Year 2660*, was Gernsback's first and best-known book. Clumsy in execution, it

ABOVE Luxembourgish-American journalist, editor, inventor, author and sci-fi pioneer Hugo Gernsback.

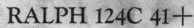

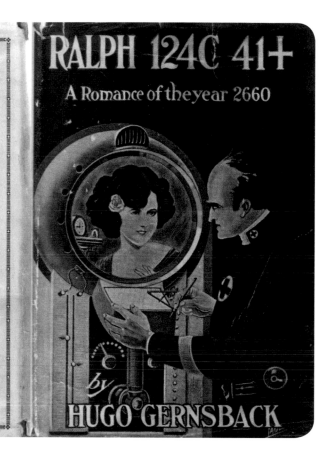

ABOVE Front and back cover featuring a Telephot for Gernsback's novel, compared by his publishers to the work of Jules Verne.

nevertheless marked an important milestone in sci-fi, both for its long list of predictions and as a flagship for the "scientific romance" approach to sci-fi: using narrative conventions to get across ideas about science and potential technological advances.

Gernsback's Telephot is often hailed as the forerunner of television, a foretaste of a technology that would not be fully realized until the 1930s. However, all of the key technologies needed for television had already been invented by the time Gernsback wrote about Ralph, and British inventor John Logie Baird achieved the first successful synthesis of these elements, producing a crude but working television broadcast, in 1926. Through his

editorship of popular science and technology journals, Gernsback was likely well informed about the myriad developments in television technology, which flourished around this period under a wonderful profusion of names, including but not limited to Distant Electric Vision, Phototelegraphy, The Electric Telescope, Telectroscopy, Hear-Seeing, Audiovision, Radio Kinema, Radioscope, Lustreer, Farscope, Optiphone and Mirascope.

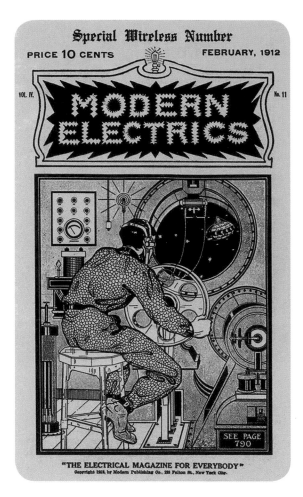

Special Wireless Number

PRICE 10 CENTS   FEBRUARY, 1912

VOL. IV.   No. 11

MODERN ELECTRICS

SEE PAGE 790

"THE ELECTRICAL MAGAZINE FOR EVERYBODY"
Copyright 1912, by Modern Publishing Co., 233 Fulton St., New York City.

LEFT One of the many Gernsback-edited and published journals exploring new science and technology.

OPPOSITE From an 1880 issue of *Scientific American,* an article exploring the technology of the selenium camera; selenium is an element that changes conductivity in response to light.

TALKING PICTURES

If the Telephot predicts anything, however, it is not television but the videophone: telecommunication with pictures as well as sound. In some respects, this is not a difficult technology to conceive; fantasies about achieving this power, through means natural or supernatural, are at least as old as stories of magic mirrors and scrying stones. Specific technological iterations of televisual communication date back at least as far as 1772, with French dramatist Louis-Sébastien Mercier predicting that Parisians of the year 2440 would be able to avail themselves of the services of "optical cabinets", in his book *L'An 2440, rêve s'il en*

fut jamais, translated into English as *Memoirs of the Year Two Thousand Five Hundred* [sic].

The advent of telegraphy and electricity raised hopes that such technology might soon be realized. In December 1878, for example, French writer and cartoonist George du Maurier, drawing in *Punch* magazine, suggested that Thomas Edison was on the verge of announcing a "telephonoscope", while in *Scientific American* in 1880, inventor George Carey penned an article outlining principles for "seeing by electricity". Just two years before Gernsback's tale, E. M. Forster's visionary 1909 short story "The Machine Stops", concerns humans confined to individual subterranean cells where they communicate only by videophone.

HOPELESS MONSTROSITIES

Despite the vision of Gernsback and others, it would be a long time before the videophone became an everyday reality – though this was not for want of trying. Just two years after Gernsback's story was published as a novel, AT&T in America managed to cobble together the video technology to create a crude but working videophone. On 7 April 1927, then commerce secretary (later president) Herbert Hoover spoke by videophone with Bell Labs in New York. In a 2004 article in the journal *Technological Forecasting and Social Change,* "On the Persistence of Lackluster Demand – the History of the Video Telephone", Steve Schnaars and Cliff Wymbs describe this early videophone as:

JUNE 5, 1880.]

Scientific American.

SEEING BY ELECTRICITY.

The art of transmitting images by means of electric currents is now in about the same state of advancement that the art of transmitting speech by telephone had attained in 1876, and it remains to be seen whether it will develop as rapidly and successfully as the art of telephony. Professor Bell's announcement that he had filed at the Franklin Institute a sealed description of a method of "seeing by telegraph" brings to mind an invention for a similar purpose, submitted

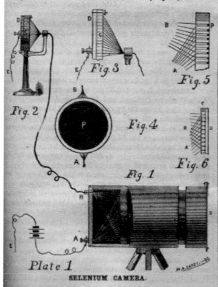

Plate 1

SELENIUM CAMERA.

to us some months since by the inventor, Mr. Geo. R. Carey, of the Surveyor's Office, City Hall, Boston, Mass. By consent of Mr. Carey we present herewith engravings and descriptions of his wonderful instruments.

Figs. 1 and 2, Plate 1, are instruments for transmitting and recording at long distances, permanently or otherwise, by means of electricity, the picture of any object that may be projected by the lens of camera, Fig. 1, upon its disk, P. The operation of this device depends upon the changes in electrical conductivity produced by the action of light in the metalloid selenium. The disk, P, is drilled through perpendicularly to its face, with numerous small holes, each of which is filled partly or entirely with selenium, the selenium forming part of an electrical circuit.

The wires from the disk, P, are insulated and are wound into a cable after leaving binding screw, B. These wires pass through disk, C (Fig. 2), in the receiving instrument at a distant point, and are arranged in the same relative position as in disk, P (Fig. 1).

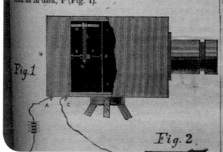

Fig. 2

the electrical current, thereby giving a luminous image instead of printing the same. These platinum or carbon points are arranged relatively the same as the selenium points in Plate P (Figs. 1 and 4); each platinum or carbon point is connected with one of the wires from selenium point in disk, P (Fig. 1), and forms part of an electrical circuit.

The operation of the apparatus is as follows: If a white letter, A, upon a black ground be projected upon disk, P (Fig. 1), all parts of disk will be *dark*, excepting where the letter, A, is, when it will be light; and the selenium points in the light will allow the electric current to pass, and if the wires leading from disk, P (Fig. 1), are arranged in the same relative position when passing through disk, C (Fig. 2), the electricity will print upon the chemically prepared paper between C and D (Fig. 2), a copy of the letter, A, as projected upon disk, P (Fig. 1). By this means any object so projected and so transmitted will be reproduced in a manner similar to that by which the letter, A, was reproduced.

Figs. 1 and 2, Plate 2, are instruments for transmitting and recording by means of electricity the picture of any object that may be projected upon the glass plate at T T (Fig. 1), by the camera lens. The operation of these instruments depends upon the changes in electrical conductivity produced by the action of light on the metalloid selenium.

The clock-work revolves the shaft, K, causing the arm, L, and wheel, M, to describe a circle of revolution. The screw, N, being fastened firmly to wheel, M, turns as wheel, M, revolves on its axis, thus drawing the sliding piece, P, and selenium point, disk, or ring, B, towards the wheel, M—see Fig. 3. These two motions cause the point, disk, or ring, B, to describe a spiral line upon the glass, T T, thus passing over every part of the picture projected upon glass, T T.

The selenium point, disk, or ring will allow the electrical current to flow through it in proportion to the intensity of the lights and shades of the picture projected upon glass plate, T T.

The electric currents enter camera at A, and pass directly to the selenium point, disk, or ring, B; thence through the sliding piece, P, and shaft, K, by an insulated wire to binding screw, C (Fig. 1); from this screw by wire to binding screw, D (Fig. 2), through shaft, K, and sliding piece, P, to point, E (Fig. 2); then through the chemically prepared paper placed against the inner surface of the metallic plate, X X, by wire, F, to the ground, thus completing the circuit and leaving upon the above mentioned chemically prepared paper an image or permanent impression of any object projected upon the glass plate, T T, by the camera lens.

Fig. 2 is the receiving instrument, which has a clock movement similar to that of Fig. 1, with the exception of the metallic point, E, in place of the selenium point, disk, or ring (Fig. 1), at B.

Fig. 3 is an enlarged view of clock-work and machinery shown in Figs. 1 and 2.

Oil in Allegany County, New York.

The Albany *Journal*, of April 22, reports that oil in paying quantities is being developed near Wellsville, in Allegany County, about forty miles to the northeast of what is known as the Bradford district in Pennsylvania. On Monday, April 19, an undoubted forty-barrel well was struck at a point less than three miles from Wellsville. It is near the Triangle Well, which has been flowing moderately for two or three months, and about six miles from the Pennsylvania line. The event causes great excitement in that locality, as the fact is now placed beyond doubt that the Bradford belt, as it is called, extends indefinitely in a northeasterly direction into New York State. The region between Olean and Wellsville is now in fair way of being developed into first class oil territory.

NOVEL ANIMAL MOTOR.

Animals have always been used as a source of motive power, but the machinery for utilizing this power has generally been of such clumsy and imperfect construction that

… a hopeless monstrosity that took up half a room. It had two large screens – a small screen that provided excellent picture quality but was too small to see and a larger screen that offered little more than a crude silhouette. Technically speaking, the screen flashed 18 frames per second, fast enough to seem animated.

ABOVE Alexander Graham Bell's photo phone used a selenium semi-conductor to detect light pulses triggered by acoustic vibrations.

OPPOSITE A 1927 imagining of a videophone system; perhaps more properly, a video and a phone.

Schnaars and Wymbs argue that the case of the videophone illustrates how, while "some radically new technological products soar smoothly from introduction to stunning market growth, just as textbooks say they should … that is not always the case, nor is it even the most likely outcome". Writers like Gernsback would typically fail to anticipate the mundane and frustrating barriers to the rapid development and widespread adoption of radical technology, although they could imagine possible form factors for such technologies. On this last point, however, it is interesting to note the phenomenon sometimes known as Zeerust, a

neologism borrowed from a place name by Douglas Adams and John Lloyd, in their 1983 lexicon of necessary but non-existent words, *The Meaning of Liff*, meaning "The particular kind of datedness which afflicts things that were originally designed to look futuristic". Contemporary or near-contemporary illustrations of the Telephot, for instance, typically show a large, bulbous cabinet, conforming to contemporary notions of, for example, cathode-ray tube technology.

The Germans were the next to attempt to realize videophone technology. In 1936 videophones, known as *Gegensehn-Freisprechanlage* ("visual telephone

LE PETIT INVENTEUR

Albin MICHEL
ÉDITEUR
22, rue Huyghens, 22
PARIS (14ᵉ)

ABONNEMENTS :
FRANCE...... **12** francs
ÉTRANGER.. **18** francs

LA TÉLÉVISION

Tubes de Néon

Moteur de Synchronisme

Courant

Bientôt on verra la personne à qui l'on parle d'aussi loin qu'on l'entend au téléphone.

system"), for public use were introduced in Germany. Invented by a German post office engineer, Dr Georg Schubert, the service linked Berlin with Leipzig, 150 kilometres (93 miles) away. It was expensive and did not prove popular, and the service was interrupted by the war and ceased in 1940.

A similar experiment in America also failed. In 1964 AT&T introduced the Picturephone, video calling via stations in booths installed in New York, Chicago and Washington. The company's in-house magazine confidently predicted that the Picturephone would "displace today's means of communication, and in addition will make many of today's trips unnecessary".

But the calls were expensive ($16–$27 – around $100–$200 in today's money – for three minutes) and limited: it was necessary to reserve a spot, and of course a call could only be placed to another booth. The service was shut down in 1968.

A number of hurdles had to be cleared before Gernsback's vision could become routine reality. Video calling had to become cheap or free, and the equipment needed had to become ubiquitous enough to allow a useful number of potential contacts. In order to fulfil these requirements video calling had to wait on other developments, such as market penetration of cheap internet-connected cameras.

FROM TELEVISION TO TELEPRESENCE

What can science fiction tell us about likely future developments in video calling? Holograms in place of two-dimensional screen projections are commonplace in sci-fi, but it is not clear that in the real world the complexity of achieving this would be justified by demand. More likely avenues are perhaps suggested by the overlap between video calling and other themes in sci-fi. Science-fiction treatments of bionics and cyborgs (see Chapter 15) have encouraged real researchers to explore ways to connect the human body to the telecommunications network. A pioneer in this field is Kevin Warwick, an engineering professor whose "Project Cyborg" involves using himself as a guinea pig. Warwick, who explicitly draws his inspiration from sci-fi, has had wirelessly communicating electronic chips implanted in his body, enabling him to connect to remote devices such as robot arms.

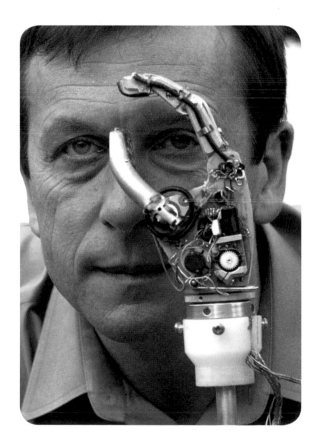

OPPOSITE Advert for AT&T's 1964 Picturephone, playing up the technology's combination of telephone and television.

ABOVE A public videophone from the 1982 film *Blade Runner,* however futuristic the designers' vision, they failed to predict that mobile phones would make public ones obsolete.

RIGHT Professor Kevin Warwick and the robot hand he can control remotely using implanted electronics.

Such connections between individual nervous systems and the digital world open new worlds of telepresence, the logical extension of which will be telepresently embodied consciousness. Telepresent control of robot avatars in fiction dates back at least as far as Joseph Schlossel's 1928 story "To the Moon by Proxy", in which remotely operated robots explore the Moon. In the following year Japanese writer Haruo Sato proposed in his story "Nonsharan Kiroku" ("Nonchalant Record") an advanced communication system that could interactively transmit sensory information, including touch. The real-world science of telepresence traces its origins to a 1980 article, "Telepresence", by Marvin Minsky, who cited a 1942 sci-fi novel as inspiration: "My first vision of a remote-controlled economy came from Robert A. Heinlein's prophetic novel, *Waldo*."

One extreme extension of telepresence – or perhaps telexistence is a better term – is represented by the central conceit of the 2009 James Cameron film *Avatar*. In this movie humans are able to transfer their conscious minds into alien bodies, in order to explore and interact with an otherwise hostile world. While this seems a long way from a video call, it illustrates the possibilities of a future in which it is routine to be telepresent in a distant location through either a virtual or robot avatar, and thus to achieve the ultimate in telecommunication.

ABOVE Promotional image for the film *Avatar*, showing the lead character and the specially grown alien body he uses for "telexistence".

OPPOSITE Paperback edition of *Waldo*, a novel by prolific sci-fi legend Robert Heinlein that is often cited for its prescient treatment of automation, robotics and telepresence.

17

TABLETS AND TRICORDERS:
THE PORTABLE TECH INSPIRED BY *TREK*

Science and technology sometimes progresses in mighty leaps, but more generally it advances incrementally, contingent on prior research.

Even major discoveries and inventions usually have to wait for the correct conditions and precursors: each advance expands the territory of the "adjacent possible". The development of digital technology is a good example. Some of the principles had existed since the Victorian era, with the work of George Boole and Charles Babbage on logic and logic engines; others since the 1940s, with the work of Alan Turing. But the explosive advances in digital electronics from the 1970s were contingent on the development of the transistor and integrated circuits.

Thus it was not until the 1970s that personal digital technology development could really take off, and this in turn meant that this technology would be shaped by the media that had inspired the generation of engineers and entrepreneurs who came of age in the 1970s, in particular the popular screen science fiction of the 1960s.

Above all, this meant two works: *Star Trek* and *2001: A Space Odyssey*. This TV show and film both featured gadgetry and devices that have since become iconic, and which have been directly responsible for inspiring the gadgetry and devices that have come to dominate modern personal technology – including the smartphone and the tablet computer – and perhaps soon to include portable, personal medical devices.

YOUR MOST VALUABLE SINGLE POSSESSION

Portable connected tech devices – often known in sci-fi as personal digital assistants (PDAs) – were predicted before *Star Trek*. For example, in Robert Heinlein's 1948 novel *Space Cadet*, everyone is equipped with pocket-sized mobile communication devices. Perhaps the most dramatically foresighted

incarnation of PDAs in all of science fiction dates to 1965, the year before *Star Trek* was first broadcast. In Frederik Pohl's tale *The Age of the Pussyfoot*, a sleeper awakes in the future and is briefed on what his guide calls "your most valuable single possession in your new life: [a] remote-access computer transponder called the 'joymaker'". This remarkable device is "a combination of telephone, credit card, alarm clock … reference library, and full-

time secretary". Pohl's joymakers have a range of features that are familiar from modern smartphones: voice recognition software, voicemail and medical diagnostic apps; and a few that are not yet quite standard, such as haptic VR for virtual kissing. Pohl even painted a chillingly prescient picture of how such devices would harvest personal data that cloud computing services would use for personal profiling, for instance in the following exchange:

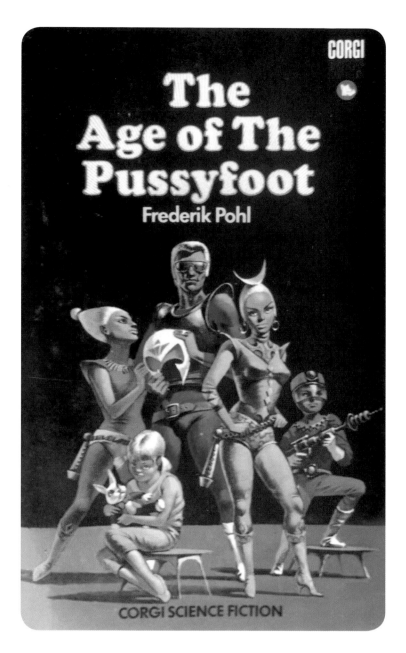

LEFT Cover art for Frederik Pohl's *The Age of the Pussyfoot*, a "sleeper awakes" tale featuring a device remarkably like a modern smartphone.

OPPOSITE LEFT A classic, clamshell-style communicator from the original 1967 TV series *Star Trek*.

OPPOSITE RIGHT British sci-fi colossus Arthur C. Clarke, who came up with the Newspad.

"Have you filled out an interests profile?" "I don't think so." "Oh, do! Then it will tell you what programs are on, what parties you will be welcomed at, who you would wish to know … Let the joymaker help you." "…I don't understand," he said. "You mean I should let the joymaker decide what I'm going to do for fun?" "Of course. There's so much. How could you know what you would like?"

Despite the amazing accuracy of his predictions, Pohl's joymaker is little remembered today. Instead, the best-known science fiction progenitor of the mobile phone is the communicator device featured in *Star Trek*, an elegant piece of production design that featured a clamshell form-factor and a characteristic alert sound, both of which have been carefully replicated by some modern mobile phones.

FANTASTIC TECHNOLOGY

Perhaps the clearest example of a modern PDA being inspired by a science-fiction device is the touchscreen tablet computer. The best known incarnation of the tablet is Apple's iPad, which has strong links to the similarly named *Star Trek* device known as the Personal Access Display Device or PADD (see page 207). Samsung, on the other hand, in the course of a protracted and expensive legal battle with Apple over the origins and inspirations

for the tablet, pointed to that other pillar of 1960s screen sci-fi: *2001: A Space Odyssey*. Defending themselves against accusations of having plagiarized tablet design concepts from Apple, Samsung cited the classic sci-fi film.

Stanley Kubrick's 1968 epic masterpiece, based on a short story by Arthur C. Clarke and concocted in collaboration with the British sci-fi legend, features a device called a Newspad. It was by no means the first piece of sci-fi to show newspapers being consumed via screens; that distinction may belong to the 1890 novel *Caesar's Column: A Story of the Twentieth Century*, by American writer, politician and mythmaker Ignatius Donnelly, in which content such as menus and newspapers are presented on screens. But the Newspad looked, worked and was used in ways that will seem extremely familiar to any modern tablet user. In his novelization of the movie, Clarke described the Newspad and its operation at length:

2001: a space odyssey

MGM PRESENTS A STANLEY KUBRICK PRODUCTION

CINERAMA® Super Panavision® and Metrocolor

When he tired of official reports ... he would plug his foolscap-sized Newspad into the ship's information circuit and scan the latest reports from Earth. One by one he would conjure up the world's major electronic papers ... When he had finished [an article], he would flash back to the [front] page and select a new subject for detailed examination.

Clarke has his protagonist – a space-going NASA scientist named Heywood Floyd – reflect on the futuristic nature of the Newspad:

Floyd sometimes wondered if the Newspad, and the fantastic technology behind it, was the last word in man's quest for perfect communications. Here he was, far out in space, speeding away from Earth at thousands of miles an hour, yet in a few milliseconds he could see the headlines of any newspaper he pleased. (That very word "newspaper," of course, was an anachronistic hangover into the age of electronics.)

There is even room for a very modern-sounding rumination: "one could spend an entire lifetime doing nothing but absorbing the ever-changing flow of information from the news satellites". The movie's production designers took advice from IBM (the Newspad was intended to be an IBM product, reflecting the one-time dominance of the American company in the field of information technology), and created a device so reminiscent of a modern tablet that Samsung cited it as an example of "prior art" in their defence against an Apple lawsuit claiming patent infringement.

ABOVE Nichelle Nichols as Lieutenant Uhura in *Star Trek*, using an early, rather boxy iteration of the PADD.

OPPOSITE Poster for the 1968 film *2001: A Space Odyssey*, showing lunar astronauts equipped with some sort of tablet computer.

PADDS, PODS AND PADS

Samsung might also have cited the production design of the 1987–1994 TV series *Star Trek: The Next Generation* (*TNG*) as an example of prior art. This series, and subsequent spin-offs, featured a variety of tablet computers, most of which were models of Starfleet's Personal Access Display Device or PADD. The typical PADD was a flat, rectangular handheld device about 15 centimetres (6 inches) high, consisting mostly of screen, which could be operated by touch or stylus. By linking to the ship's computer, when on a starship, the PADD could be used to access most of the computer's functions, allowing it to serve as a portable workstation, clipboard, notepad and computer screen. It was supposed to represent a technological evolution of a much bulkier electronic clipboard seen on screen in the original series, which looked more like a small typewriter than a tablet computer.

Most of the *TNG* PADDs were designed by Rick Sternbach, who explained in an interview in *Star Trek: The Magazine*, "I always assumed that the PADD would be a highly capable device, able to communicate with other tech devices. The fact that we have devices like it today doesn't surprise me in the least. They're all very, very cool, but I expected them to show up eventually." Many have commented on the apparent similarity between the PADD and the iPad. The actor Brent Spiner, who played the android Data in the *TNG* series, has said he sees a very direct connection between the two, but Sternbach himself sees the link in a wider context: "I can understand why there's been some hoopla over the comparison to recent tablet computers, particularly the Apple iPad, but I really see the PADD as simply an outgrowth of science fiction data displays imagined for decades in literature and on screen."

In fact, Steve Jobs was an admitted "Trekkie" (a fan of *Star Trek*), and said that *Trek* tech such as the PADD did indeed inspire some of his company's designs, although he was talking about the precursors to the iPad, particularly the iPod. The iPod is a portable digital music player, and Jobs had described how one inspiration for this device was seeing an episode of *TNG* in which Data is using touch technology to navigate a menu of music.

A REMARKABLE MINIATURIZED DEVICE

Sternbach based his concept of the PADD on the original *Star Trek* portable tech device: the tricorder. This is imagined as a portable device equipped with significant computing power, which may be networked, although this was not originally explicit. The main point of difference between the tricorder and a PDA is that the former has a relatively restricted set of functions,

LEFT DeForest Kelley as "Bones" McCoy in *Star Trek,* with his "man-bag"-style medical tricorder.

OPPOSITE TOP Steve Jobs introduces new models of the iPad, the Apple tablet that many believe to have been at least partially inspired by the *Star Trek* PADD.

OPPOSITE BOTTOM Bill Gates in 2002, demonstrating an early, commercially unsuccessful tablet computer.

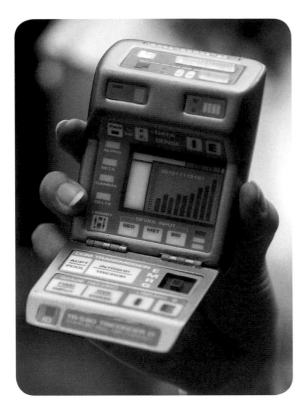

LEFT A tricorder from *Star Trek TNG*. An interesting example of Zeerust, in that it features a lot of fixed buttons and not much screen, whereas a more likely future configuration would opt for greater adaptability by having more screen.

OPPOSITE LEFT The DxtER system in action.

OPPOSITE RIGHT Scanadu founder Walter de Brouwer demonstrates the companion app for his tricorder-emulating system.

since it is used only for medical and scientific purposes. In the writers' guide for the original series, the tricorder is described as "A portable sensor-computer-recorder, about the size of a large rectangular handbag, carried by an over-the-shoulder strap. A remarkable miniaturized device, it can be used to analyze and keep records of almost any type of data on planet surfaces, plus sensing or identifying various objects." The name of the device reflected that it had three main functions: sensing, analyzing and recording. The version that became particularly well known was the medical tricorder, used by the medical personnel for scanning and diagnostics. The tricorder shrank over time to become a handheld device.

The *Trek* tricorder was not the first sci-fi iteration of a portable, accessible diagnostic device. In C. M. Kornbluth's 1950 story "The Little Black Bag", a subordinate class of humans is given simple technology that anyone can operate, including a basic but comprehensive diagnostic and medical kit. This scenario is reminiscent of today's drive to make accessible defibrillator kits that anyone can use, which come with automated instructions. As Kornbluth observed in the preamble to his tale, "The normal progress of a technology produces simpler and simpler gadgets involving more and more complex fundamental laws. And, of course, requiring less and less of the user ..."

Kornbluth's device is little known, however, and it was the tricorder that quickly became an iconic piece of sci-fi kit and an inspiration to designers in the real world. Indeed, it lent its name to a specific iteration of the XPRIZE challenges founded by Greek-American entrepreneur and futurist Peter Diamandis. Diamandis's XPRIZE Foundation seeks to jump-start progress in various fields by putting up prizes for research and inventions that meet certain goals or benchmarks.

In 2012, with sponsorship from chip manufacturer Qualcomm, the Foundation launched the Qualcomm Tricorder XPRIZE. Its remit was "to incentivize the development of innovative technologies capable of accurately diagnosing a set of 13 medical conditions independent of a healthcare professional or facility, [with the] ability to continuously measure five vital signs, and [give] a positive consumer experience".

In 2017 the Tricorder XPRIZE was awarded to a device named DxtER, created by an American team, Final Frontier Medical Devices, and led by brothers Basil and George Harris, founders of Basil Leaf Technologies. Admittedly, DxtER little resembles a Trek tricorder; the actual physical equipment involved includes various small sensors, which collect data that is analyzed along with a patient's medical history. The real innovation that, in the company's own words, lies "at the heart of DxtER is a sophisticated diagnostic engine". Basil Leaf

developed algorithms for 34 health conditions, including diabetes, sleep apnoea, tuberculosis and pneumonia.

DxtER is just one of many "real tricorders" that have been developed over the years, including devices such as portable mass spectrometers (a device for identifying the composition of substances); lab-on-a-chip gadgets, where microscopic channels etched into a silicon wafer guide samples through various tiny sensors and detectors; and a machine for measuring pressure, electromagnetic fields and temperature. The rise of the smartphone has provided an obvious platform for tricorder-style functionality, and there have been various apps and smartphone add-ons that claim to achieve this, such as the short-lived Scanadu Scout (a hockey-puck-sized adjunct to a smartphone) or Google's own Science Journal app. While a fully functioning tricorder may not yet have been achieved, it seems unlikely that we will have to wait until the twenty-third century.

18

CYBERSPACE:
FROM CARSON CIRCUITS TO THE INTERNET

Failure to predict the coming of the internet and its wide-ranging and profound impact on global culture, economy and society is often viewed as one of sci-fi's greatest missteps.

Some areas, like telecommunications or miniaturization in general, sci-fi predicted in detail; others, like robots or personal mobility, sci-fi wildly overhyped. IT in general, and the internet in particular, on the other hand, were not much explored by science fiction until they had already arrived or were imminent. But there were exceptions, one of the most eye-opening of which was a story from 1946, which predicted that every home would have a computer terminal that connected to a network of information technology, which in turn would make unprecedented amounts of information universally available, with darkly transformative consequences for individuals and society.

VERY CONVENIENT
This story was "A Logic Named Joe", by Murray Leinster (one of the pen names of American author William Fitzgerald Jenkins), first published in March 1946 in the pulp magazine *Astounding Science Fiction*. In the future world of the tale, "logics" are computer terminals. "You got a logic in your house," the narrator of the tale, a logic repairman, tells the reader, "It looks like a vision receiver used to, only it's got keys instead of dials, and you punch the keys for what you wanna get." Written at a time when computers were huge, room-filling contraptions, and experts confidently predicted that fewer than a dozen would be needed to meet global demand, this represents a high degree of prescience.

Thirty years later, personal tech pioneers such as Steve Jobs would be lauded as geniuses for anticipating a market for home computers.

Logics connect to "the Tank ... a big buildin' full of all the facts in creation and all the recorded telecasts that ever was made". The Tank in turn has "the Carson Circuit all fixed up with relays", whatever that means, and is "hooked in with all the other Tanks all over the country – an' everything you wanna know or see or hear, you punch for it an' you get it. Very convenient." This makes a decent description of the modern internet, recognized as such by the American Computer History Museum, which has called the story "one of the most prescient views of the capabilities of computers in a network".

But Leinster did not stop there. The titular Joe is a logic that achieves enough self-awareness to decide to improve its own performance – and that of the whole network – in its core task of presenting users with information that they need. As a result, logics all around the country start passing on to their users information about everything from how to cure a hangover to how to poison a spouse and get away with it or plan the perfect bank robbery. As the narrator reflects, "Joe ain't vicious, you understand. He ain't like one of those ... robots you read about that make up their minds the human race is inefficient and has got to be wiped out an' replaced by thinkin' machines. Joe's just got ambition. If you were a machine you'd wanna work right, wouldn't you? That's Joe. He wants to work right."

This plot development mirrors what many artificial intelligence boosters hope to see in the future of AI:

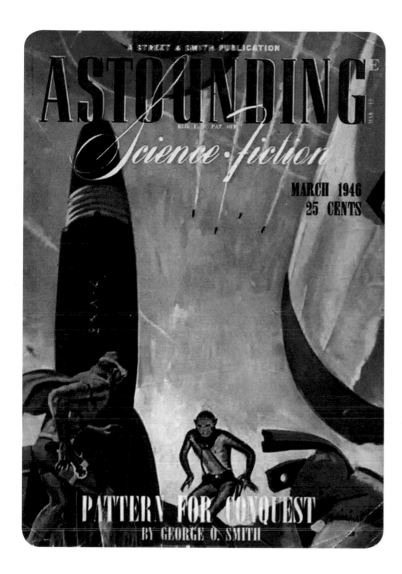

LEFT A conventional-looking space opera scores the cover of the March 1946 issue of *Astounding Science Fiction*, despite the presence within of one of the most celebrated examples of prescient sci-fi, Murray Leinster's "A Logic Named Joe".

once computers achieve intelligence, they will start to improve their own design and performance, resulting in exponential, snowballing gains in intelligence and performance that will quickly lead to godlike powers. This scenario is known as the "Singularity". But it is not necessary for networked IT to reach Singularity status in order for it to have a profound impact on humanity; the simple fact of such a network already achieves this, and Leinster foresaw this, too: "Logics changed civilization. Logics *are* civilization! If we shut off logics, we go back to a kind of civilization we have forgotten how to run!"

In the story, Joe's excessive "ambition" leads to the relatively sanitized existing version of the internet becoming something much closer to modern-day reality: a double-edged sword that can be put to malign uses. Leinster foresaw that this kind of power would serve to unleash the basest instincts of the human race. Dr Ellen Peel, Professor of English and Comparative Literature at San Francisco State University, notes that in the story "Ordinary logics already grant many wishes, but what

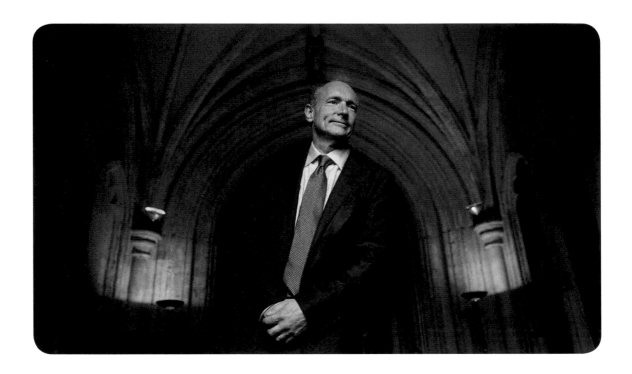

Joe grants are the forbidden ones, the 'monsters from the id'". The tale is resolved when the repairman locates Joe, switches off the logic and stores it in a cellar. "It is suggestive," notes Peel, "that the narrator stores Joe in the basement, sort of a subconscious." Of course, the real internet is more of a Pandora's box than a logic named Joe.

For all its startling accuracy and foresight, there is no evidence that "A Logic Named Joe" directly inspired the creation of the internet, which dates back to the US Department of Defense's Advanced Research Projects Agency's work on networking computers in the 1960s and 70s. This eventually led to the creation of ARPANET, the first seed of what would eventually become the internet. But science fiction *has* been referenced as a direct inspiration by Tim Berners-Lee, the British computer scientist behind the creation of the World Wide Web (aka the Web or WWW), a global hyperlinked information system – a kind of information space accessible via the internet, which would prove to be the internet's "killer app"; the

functionality that made it indispensable. Berners-Lee says that one of his inspirations was the 1964 Arthur C. Clarke short story "Dial F for Frankenstein", in which the world's telecoms networks are linked together and the resulting hyperconnected web of IT spontaneously achieves sentience and resists attempts to shut it down. This plot is probably more familiar as the premise for James Cameron's *Terminator* tales, in which a similar link-up between complex systems results in the creation of Skynet, a self-aware AI that immediately sets about wiping out humanity.

CYBERPUNK

The advent of the internet helped give birth to a new cultural movement, known as cyberpunk. Many of the tropes and markers of this culture were laid down or initially propagated by the cyberpunk sci-fi genre, created by the American-Canadian writer William Gibson and others. In his 1982 short story "Burning Chrome", Gibson coined the term "cyberspace" – several years before the genesis of the World Wide

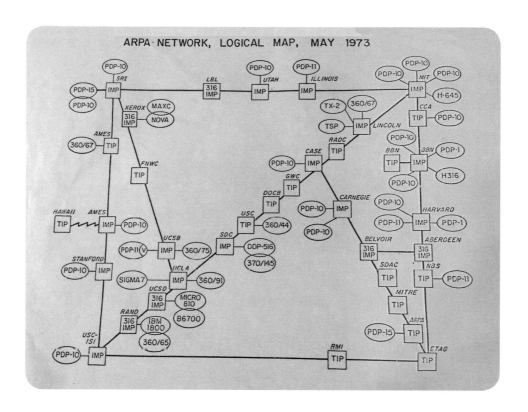

ARPA·NETWORK, LOGICAL MAP, MAY 1973

ABOVE Network map of the original 1973 ARPAnet, progenitor of the internet.

LEFT With this humble NeXT computer, Tim Berners-Lee created the earliest iterations of the World Wide Web.

OPPOSITE Tim Berners-Lee, father of the World Wide Web, who cites an Arthur C. Clarke story as one of the inspirations for his innovation.

FOLLOWING PAGES Scene from the 2009 film *Terminator Salvation*, in which Christian Bale leads the human resistance fighting the implacable Skynet.

Web opened general access to this digital terrain. Key themes in cyberpunk include hacking, immersion in online culture and online identities, and manipulation of and reaction to the way that the internet enables and mediates data gathering, information control and commodification and erosion of privacy.

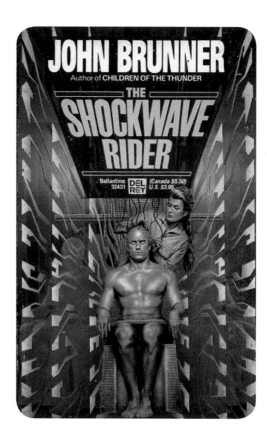

For all of Gibson's trail-blazing, however, these themes and the cyberpunk genre were anticipated in an earlier work: John Brunner's 1975 novel *The Shockwave Rider*. In this pioneering story, malign corporations exert repressive power through their control of data, and a cyberpunk rebel uses computer hacking to rebel against the system. Brunner's tale even has "worms": virus-like computer programs that can tunnel through defences and propagate in cyberspace. At the climax of the novel a worm is released that blasts open online databases in an act of radical data transparency, revealing the corrupt secrets of the corporate-state establishment.

Another facet of internet culture explored and predicted by some quite early sci-fi is its addictive and immersive nature, and the alienation and isolation from real human society resulting from total absorption into online worlds.

The ultimate extrapolation of this is fully immersive virtual reality, of the sort featured in the 1999 film *The Matrix*, in which humans are confined in pods and literally plugged into a vast virtual reality simulation via bionic implants. A much less well-known sci-fi work prefiguring very closely *The Matrix*'s world of pod people is James E. Gunn's 1961 novel *The Joy Makers*, in which a supercomputer houses human dependents in individual cells where they enjoy a state of foetal bliss, and those who reject this choice go to live on Venus. This dynamic recalls the choice advanced by the "Experience Machine" thought experiment (see page 179), which asks whether people would willingly choose to plug themselves into a virtual reality generating machine that would supply the illusion of pleasurable experience. In Gunn's novel, those who reject plugging in have to move to Venus. In *The Matrix*, rejecting the computer means scratching

out a desperate, hunted existence in a post-apocalyptic dystopia, an option so unappealing that one character actually demands to be plugged back in. Stories such as these were prefigured by the E. M. Forster 1909 short story "The Machine Stops", in which humans are confined to individual subterranean cells where they are maintained by the Machine and communicate only by videophone – a far-sighted prediction of the burgeoning reality of our atomized, antisocial society, in which people pay attention only to their screens, shutting themselves off from the real world.

OPPOSITE LEFT Cyberpunk pioneer William Gibson, author of "Burning Chrome" and *Neuromancer*.

OPPOSITE RIGHT Cover art for John Brunner's 1975 novel *The Shockwave Rider,* which anticipated many themes of cyberpunk.

ABOVE Scene from *The Matrix* (1999), in which Neo awakens from his virtual world to find the Machines have been keeping him and billions of others as "brains in vats".

FURTHER READING

GENERAL

Brian Ash, *The Visual Encyclopedia of Science Fiction*, 1977.

Brian M. Stableford, *Science Fact and Science Fiction: An Encyclopedia*, 2006.

The Encyclopedia of Science Fiction: sf-encyclopedia.com

Technovelgy: technovelgy.com

1 BURST UPON THE WORLD

Eric Ambler, *The Dark Frontier*, 1936.

Bernard and Fawn Brodie, *From Crossbow to H-Bomb*, 1973.

Cleve Cartmill, "Deadline", 1944.

Robert Cromie, *The Crack of Doom*, 1895.

Lester del Ray, "Nerves", 1942.

Robert Heinlein, "Blowups Happen", 1940.

Anson MacDonald, "Solution Unsatisfactory", 1941.

Richard Rhodes, *The Making of the Atomic Bomb*, 1986.

H. G. Wells, *The World Set Free*, 1914.

2 TANKS

Joel Levy, *50 Weapons That Changed the Course of History*, 2014.

Richard Ogorkiewicz, *Tanks: 100 Years of Evolution*, 2018.

H. G. Wells, "The Land Ironclads", 1903.

3 ENERGY WEAPONS

Margaret Cheney, *Tesla: Man Out of Time*, 1981.

Hugo Gernsback, Grant Wythoff (Ed.), *The Perversity of Things: Hugo Gernsback on Media, Tinkering, and Scientifiction*, 2016.

David Hambling, *Weapons Grade: How Modern Warfare Gave Birth to Our High-Tech World*, 2006.

H. G. Wells, *The War of the Worlds*, 1897.

4 DRONES AND AUTONOMOUS WEAPONS

Medea Benjamin, *Drone Warfare: Killing by Remote Control*, 2012.

Marc Seifer, *Wizard: The Life and Times of Nikola Tesla*, 1996.

5 CREDIT CARDS

Brian Aldiss, "The Underprivileged", 1963.

Edward Bellamy, *Looking Backward*, 1888.

Jack Weatherford, *The History of Money*, 1997.

6 BIG BROTHER IS WATCHING YOU

John Jacob Astor IV, *A Journey in Other Worlds*, 1894.

James Blish, *Cities in Flight*, 1970.

Ray Cummings, "Wandl, the Invader", 1939.

Philip K. Dick, *Lies, Inc.*, 1964.

Joel Levy, *The Little Book of Conspiracies: 50 Reasons to Be Paranoid*, 2005.

Larry Niven, *Cloak of Anarchy*, 1972.

George Orwell, *1984*, 1949.

Jack Williamson, "The Prince of Space", 1931.

Yevgeny Zamyatin, *We*, 1924.

Roger Zelazny, *This Moment of the Storm*, 1966.

7 REPLICATORS IN ACTION

Damon Knight, *The People Maker*, 1959.

Primo Levi, "Order on the Cheap", 1964.

Joel Levy, *The Infinite Tortoise: Philosophical Thought Experiments and Paradoxes*, 2016.

8 DRIVE TIME

Isaac Asimov, *The Complete Robot*, 1982.

Isaac Asimov, "Sally", 1953.

David H. Keller, "The Living Machine", 1935.

9 SUBMARINES

Iain Ballantyne, *The Deadly Trade: The Complete History of Submarine Warfare From Archimedes to the Present*, 2018.

Jules Verne, *Twenty Thousand Leagues Under the Sea*, 1870.

10 MOON ROCKETS IN FLIGHT

Cyrano de Bergerac, *Other Worlds: The Comical History of the States and Empires of the Moon and Sun*, 1657.

Andrew Chaikin, *A Man on the Moon: The Voyages of the Apollo Astronauts*, 1994

Hergé, *Tintin: Destination Moon*, 1953.

Johannes Kepler, *Somnium*, 1608.

Megan Prelinger, *Another Science Fiction: Advertising the Space Race 1957–1962*, 2010.

Jules Verne, *From the Earth to the Moon*, 1865.

H. G. Wells; *The First Men in the Moon*, 1901.

11 INTERPLANETARY TRAVEL TO MARS

Ray Bradbury, *The Martian Chronicles*, 1950.

Wernher von Braun, *Project Mars: A Technical Tale*, 1949.

Edgar Rice Burroughs, *A Princess of Mars*, 1917.

Charles Dixon, *1500 Miles an Hour*, 1895.

John Munro, *A Trip to Venus*, 1897.

12 ELECTRA'S MAGIC RAYS

Gordon Giles, "Diamond Planetoid", 1937.

Bettyann Holtzmann Kevles, *Naked to the Bone: Medical Imaging in the Twentieth Century*, 1997.

Philander, "Electra: A Physical Diagnostic Tale of the Twentieth Century", 1893.

13 ORGANISM ENGINEERING

Lois McMaster Bujold, *Falling Free*, 1988.

Matthew Cobb, *Life's Greatest Secret: The Race to Crack the Genetic Code*, 2015.

Joel Levy, *Frankenstein and the Birth of Science*, 2018.

Mary Shelley, *Frankenstein*, 1818.

H. G. Wells, *The Island of Dr. Moreau*, 1896.

14 PROZAC NATION

Aldous Huxley, *Brave New World*, 1932.

Peter D. Kramer, *Listening to Prozac: A Psychiatrist Explores Antidepressant Drugs and the Remaking of the Self*, 1993.

Norman Ohler, *Blitzed: Drugs in the Third Reich*, 2018.

15 BIONIC PEOPLE

Martin Caidin, *Cyborg*, 1972.

Raymond Z. Gallun, "Mind Over Matter", 1935.

Edgar Allan Poe, "The Man That Was Used Up", 1839.

Perley Poore Sheehan and Robert H. Davis, "Blood and Iron", 1917.

Thomas Rid, *Rise of the Machines: The Lost History of Cybernetics*, 2016.

16 VIDEO PHONES

David E. Fisher, *Tube: The Invention of Television*, 1996.

E. M. Forster, "The Machine Stops", 1909.

Hugo Gernsback, *Ralph 124C 41+: A Romance of the Year 2660*, 1925.

Robert A. Heinlein, *Waldo*, 1942.

17 TABLETS AND TRICORDERS

Michael Benson, *Space Odyssey: Stanley Kubrick, Arthur C. Clarke, and the Making of a Masterpiece*, 2018.

Arthur C. Clarke, *2001: A Space Odyssey*, 1968.

C. M. Kornbluth, "The Little Black Bag", 1950.

Frederik Pohl, *The Age of the Pussyfoot*, 1969.

18 CYBERSPACE

John Brunner, *The Shockwave Rider*, 1975.

Arthur C. Clarke, "Dial F for Frankenstein", 1964.

William Gibson, "Burning Chrome", 1982.

James E. Gunn, *The Joy Makers*, 1961.

Murray Leinster, "A Logic Named Joe", 1946.

INDEX

Page numbers in *italics* indicate illustrations

CREDITS

The publishers would like to thank the following sources for their kind permission to reproduce the pictures in this book.

Key: t = top, b = bottom, l = left, r = right & c = centre

AKG-Images: 33; /Science Source 152

Alamy: AF Archive 115, 126b, 219; /AFP 87; /Allstar Picture Library 186; /Archive 84-85; /Atlaspix 216-217; /Martin Bennett 59; /Ian Bottle 73; /Raymond Boyd 94; /Chronicle 28b, 42, 43, 182; /dpa picture alliance 211l; /Emka74 131; /Everett Collection Inc 17, 22, 63, 89; /FLHC 1E 26, 151; /Granger Historical Picture Archive: 19, 72, 99, 162; /Historic Images 112, 168; /The History Collection 97; /Jerry Holt/Minneapoli 165; /ITAR-TASS 128; /Lebrecht Music & Arts 137r; /Moviestore Collection Ltd 36, 113; /Nearthecoast.com 200; /Photo12 125l, 185; /Pictorial Press Ltd 13; /The Print Collector 152l, 161l; /Protected Art Archive 39; /Sergi Reboredo 130; /Science History Images 38; /ScreenProd/Photononstop 55; /TCD/Prod.DB 138-139; /Trinity Mirror/Mirrorpix 184; /ullstein bild 118; /United Archives GmbH: 173; /Jim West 108-109; /Chris Wilson 205l; /WorldPhotos: 12; /World History Archive 35, 124, 127, 150

American Express: 70

Juan Carlos Izpisua Belmonte: 164

Bridgeman Images: © British Library Board. All Rights Reserved 66, 153r; /Genie: 92; /De Agostini Picture Library 37, 161r

Getty Images: AFP 87; /Keith Beaty/Toronto Star 95; /Bettmann 23, 80, 104, 121b, 123, 170, 207; /Buyenlarge 121t; /CBS 89b; /Corbis 16; /De Agostini 58; /Ed Clark/Life Magazine/The LIFE Picture Collection 18; /Kevork Djansezian 147; / Alfred Eisenstaedt/The LIFE Picture Collection 181, 192; /Thierry Falise/LightRocket 79; /Fine Art Images/Heritage Images 25; /Giorgos Georgiou/NurPhoto 102; /GraphicaArtis 107t, 191; /Maxim Grigoryev/TASS 96; /Martha Holmes/The LIFE Images Collection 179; /Paul S. Howell/Liaison Agency 176; /Hulton-Deutsch Collection/CORBIS 32; /Kurt Hutton/Picture Post 11; /Mark Kauzlarich/Bloomberg 100; / Kim Kulish/Corbis 211r; /Los Alamos National Laboratory/The LIFE Picture Collection 20-21; /Bob Farley/The Washington Post 202; /Jamie McCarthy 188; /Win McNamee 187; /Peter Macdiarmid 214; /John Moore 62; /Movie Poster Image Art 45; /MPI 15; /National Motor Museum/Heritage Images 31; /Pallava Bagla 107b; /Popperfoto 172; /Preint Collector 197; /George Rinhart/Corbis 29; /Nina Ruecker 81; /David Savill/Stringer 169; /SSPL 119, 134, 205r; /Wallace Kirkland/The LIFE Picture Collection 14; /Joe Raedle 9; /Underwood Archives 198; /Universal History Archive/UIG 196

Gonzo Bonzo via Wikimedia Commons: 218l

Jastrow via Wikimedia Commons: 163

Library of Congress, Washington: 154

Lockheed Martin: 48, 50-51

Mary Evans Picture Library: 40

McGeddon via Wikimedia Commons: 106

NASA: 133, 135, 136

Private collection: 6, 28t, 54, 56, 71, 74, 75, 93, 105, 114, 126t, 140b, 157, 175, 193, 194, 195, 203, 204, 213, 215t, 218

Rauner Special Collections Library: 155

Science Photo Library: Coneyl Jay 164; /Detlev Van Ravenswaay 143, 144, 145; /Nikola Tesla Museum 41, 57; /Joyce R. Wilson 46b

Shutterstock: 199r, 208; /American Zoetrope/Warner Bros/Kobal 174; /AP 140t; /AIP/Kobal 160; /Granger 110, 111; /20th Century Fox/Dreamworks/Kobal 86-97; /Linda R Chen/Touchstone/Kobal 82-83; /Manu Fernandez/AP 91; /Paul Grover 210; /Keystone/Zuma 122, 208t; /Lemberg Vector Studio 149; /London Films/United Artists/Kobal 7; /Lucasfilm/Bad Robot/Walt Disney Studios/Kobal/ 46t; /Greg Mathieson 189; /OLS 65; /Paramount Television/Kobal 90, 209; / Pixinoo 76; /Propstore.com 199l; /Rhimage 77; /Snap 167, 206; / Twentieth Century-Fox Film Corporation/Kobal 201; / Umbrella/Rosenblum/Virgin/Kobal 72t; /United Artists/Fantasy Films/Kobal 178; /Universal History Archive 116-117, 137; / Universal/Kobal 141; /Universal TV/Kobal 101, 177; /Igor Zhilyakov 67

SpaceX: 146-147

U.S. Air Force: 60t, 61

U.S. Navy/John F. Williams: 60b

Wellcome Collection: 183

Every effort has been made to acknowledge correctly and contact the source and/or copyright holder of each picture and Carlton Books Limited apologises for any unintentional errors or omissions, which will be corrected in future editions of this book.